The TAB™ Guide
to Vacuum
Tube Audio

About the Author

Jerry C. Whitaker is vice president for standards development at the Advanced Television Systems Committee (ATSC) in Washington, D.C. ATSC is an international standards development organization focused on advanced digital television systems. Mr. Whitaker was previously president of Technical Press, a consulting company based in the San Jose, California, area. Mr. Whitaker has been involved in various aspects of the electronics industry for over 30 years, with specialization in communications. His current book titles include the following:

- Editor-in-chief, *Standard Handbook of Video and Television Engineering*, 4th ed., McGraw-Hill, 2002
- Editor-in-chief, *Standard Handbook of Audio and Radio Engineering*, 2nd ed., McGraw-Hill, 2001
- Author, *DTV Handbook*, 4th ed., McGraw-Hill, 2006
- Editor, *Television Receivers*, McGraw-Hill, 2001

Mr. Whitaker has lectured extensively on the topic of electronic systems design, installation, and maintenance. He is the former editorial director and associate publisher of *Broadcast Engineering* and *Video Systems* magazines, and a former radio station chief engineer and television news producer.

Mr. Whitaker is a fellow of the Society of Broadcast Engineers (SBE) and an SBE-certified professional broadcast engineer. He is also a fellow of the Society of Motion Picture and Television Engineers.

The TAB™ Guide to Vacuum Tube Audio

Understanding and Building Tube Amps

Jerry C. Whitaker

New York Chicago San Francisco Lisbon
London Madrid Mexico City Milan New Delhi
San Juan Seoul Singapore Sydney Toronto

The **McGraw·Hill** Companies

Cataloging-in-Publication Data is on file with the Library of Congress

The TAB™ Guide to Vacuum Tube Audio: Understanding and Building Tube Amps

1234567890   QFR QFR   10987654321

ISBN   978-0-07-175321-0
MHID   0-07-175321-4

Sponsoring Editor
Roger Stewart

Editorial Supervisor
Janet Walden

Project Manager
Anupriya Tyagi, Cenveo
Publisher Services

Acquisitions Coordinator
Joya Anthony

Copy Editor
Lisa McCoy

Proofreader
Madhu Prasher

Indexer
Claire Splan

Production Supervisor
James Kussow

Composition
Cenveo Publisher Services

Illustration
Cenveo Publisher Services

Art Director, Cover
Jeff Weeks

Cover Design
Kelly Par

This book is dedicated to my wife, Laura, with appreciation for allowing me to indulge my hobby of building things.

Contents at a Glance

Contents

x Contents

Preface

Vacuum tubes have been around for a very long time. Appreciated for their distinctive sound, amplifiers built around tubes have found a permanent home with audio enthusiasts and experimenters alike. This book is intended for hobbyists interested in understanding tube technology and building high-fidelity audio amplifiers.

This book covers the theory and operation of vacuum tubes and audio amplifier circuits, and includes practical projects for the experienced hobbyist. The theory of tubes and amplifier circuit design is covered from a classical perspective, drawn from some of the classic work on the subject written by Karl Spangenberg as part of the McGraw-Hill *Electrical and Electronics Engineering Series* published in the mid-1940s. The idea for this book, in fact, came from a chance opportunity to acquire several books in the series.[1] Some of the theory of tube operation and basic circuits described in the projects contained in this book are adapted from another classic publication, the *RCA Receiving Tube Manual*.[2] Complementing this classical approach to amplifier design are current devices that preserve the many sonic benefits of the "tube sound."

A suite of projects is included for the reader to construct high-fidelity audio amplifiers and related equipment at a variety of levels of sophistication, power output, and construction preferences. This book also reviews the tradeoffs that engineers must make in designing an audio amplifier and helps the reader tailor particular circuits to meet their own objectives.

In each case, we begin with a basic circuit design from the *RCA Receiving Tube Manual* and build it—with additions, improvements, and modifications along the way based on experience gained from the project. This approach mirrors the techniques used by experimenters and hobbyists for decades. The goal is to produce a useful end product, learn something about the circuit, and have fun doing it.

The initial platform for each project is a test bed where variations on the basic circuits are tried and documented. The end result is a final design that is built as a showcase project. The test bed approach is very useful, since few projects turn out as expected the first time. Design, of course, is an iterative process.

---

[1] I highly recommend Terman's, *Radio Engineering* (1947) and Spangenberg's *Vacuum Tubes* (1948). These books have been long out of print; however, they are occasionally available from technical libraries.
[2] Also recommended, and also long out of print.

The audio projects presented in this book will be a finishing point for some, and a starting point for others. No doubt some readers will want to continue improving and tweaking the designs to gain the last bit of performance. Others will be content to finish the project and move on to something else.

In preparing this book, the author considered various ways of handling projects. The primary options included: 1) offer a large number of circuits with little supporting detail, or 2) offer a small number of circuits with considerable supporting detail. The author took option #2. Underlying this approach is the assumption that most readers will either be coming to vacuum tube audio projects for the first time or coming back after many years (or decades) of doing something else as a hobby (like building computers and then fighting with the software).

For audiophiles who have been building vacuum tube amplifiers for some time, there are a number of books containing interesting circuits intended for the veteran builder. In this book, we take a slower, more detailed, more methodical approach.

One of the challenges in building vacuum tube projects today is finding the parts necessary to complete the product. The author has attempted in all projects to specify parts that can be acquired readily at a reasonable cost. Anyone working on vacuum tube projects will appreciate the difficulty of finding the right part. Sometimes compromises are necessary to avoid spending considerable time and money acquiring a particular device. For these situations, the author has attempted to outline the options and tradeoffs.

The appeal of vacuum tube-based amplifiers is well known. For many enthusiasts, the challenge is largely how to make these systems understandable and to make the projects practical to build. The projects in this book include detailed schematic diagrams, layout suggestions, and parts lists. Every effort has been made to document projects that are useful, practical, and fun to build.

Like any book, this one is based on the experiences and interests of the author. Over the years there have been a number of excellent books published on the subject of audio amplifiers. Some of these books are highly technical and intended for audiophiles who are looking for peak performance from their systems. Others are hobbyist books that focus strictly on projects. This book falls somewhere in between. A concerted effort has been made to include background theory and operation of vacuum tubes. However, this book does not go into the level of detail provided by other authors who have written excellent highly technical books on the subject.[3] If you are looking for a book that will help you design an amplifier from the ground up and construct key components to achieve the last measure of performance from your system, this book is probably not what you are looking for. If, on the other hand, you have an interest in vacuum tube audio amplifiers and want to build a system as hobby, this book is probably for you.

The book includes a considerable amount of detail on fundamental principles. It is the author's belief that good design can come only from a firm understanding of the fundamentals. To that end, much of this material is developed from previous

---

[3] Notable among these is Morgan Jones and his McGraw-Hill titles *Valve Amplifiers*, 3rd ed., and *Building Valve Amplifiers*.

publications by the author (there are about 30 at last count). Readers interested in exploring the fundamental principles in greater detail are encouraged to check out the references noted in each chapter.

In the realm of audio amplifiers, a logical division can be made between audio systems intended for faithfully amplifying input signals and those intended for applications where the characteristics of the amplifier are adjusted to yield a particular "sound." A common example of the latter is the guitar amplifier. This is a specialized area of audio technology that deserves, and has, a following of its own. Many books are available that cover this technology in detail. This book is not one of them.

Fortunately, a wide variety of publications are available to audio enthusiasts and experimenters that cover a wide range of interests. In addition, vacuum tube suppliers have online web resources that provide valuable data and insight into use of specific devices. Readers are encouraged to utilize all available resources, since each has its place in the realm of education on the fascinating topic of vacuum tubes. In addition, several web-based bulletin boards are available where members discuss projects and solve problems.[4]

The objectives of this book can be distilled to the following:

- Reaffirm an appreciation for vacuum tubes
- Provide readers with a firm understanding of the operation of tubes and related components
- Build some interesting and useful audio devices
- Have fun

Readers of a certain age will likely recall the Heathkit products that were widely marketed during the 1960s and 1970s. (There were others in the space as well.) While these kits may not always be remembered for their high-end sonic audio offerings,[5] they nonetheless imparted in the hobbyists who bought them a fondness for building electronic devices and then enjoying the fruits of their labors. It is my hope that this book will recapture that spirit.

—Jerry Whitaker

---

[4] The site "diyAudio" is one such service: www.diyaudio.com/index.php
[5] There are notable exceptions, such as the WM-5 amplifier—a nice sound in any decade.

Acknowledgments

I would like to acknowledge the support and encouragement of Roger Stewart and Steve Chapman at McGraw-Hill. Without their help this book would not have been possible.

Thanks are also due to friends and colleagues Mark Richer and Rich Chernock—both of whom have a love of vacuum tubes (and are a lot smarter than me). In addition, acknowledgment is due to my brother David—an accomplished audio studio engineer—who conducted valuable listening tests on several amplifiers (and now won't give them back).

And of course, thanks to Laura for not complaining as my workbench expanded week-by-week to accommodate the boxes the UPS guy kept bringing.

Before We Get Started

Although the following text probably reads like boilerplate disclaimer language, the intent here is to clearly state some important (although perhaps obvious) points before we start heating up the soldering iron. Readers should be aware that

- Through publication of this book, the author is not rendering professional engineering services.
- The projects described in this book have been built and tested by the author, unless otherwise noted. The test results listed were obtained as stated in the text.
- The circuits provided in this book are intended for use by hobbyists and audio enthusiasts with some basic experience in electronics projects. If the reader believes that a project may be too advanced, he or she is discouraged from attempting to build it.
- While every effort has been made to ensure that all drawings, tables, and other data contained in this book are complete and accurate, no warranties can be made that the data is perfect or otherwise without error.
- All plans, drawings, circuit descriptions, and parts lists are provided as-is. No warranty is implied or stated.
- Beyond publication of this book and the advice contained therein, no other form of technical support is implied or stated.
- In an effort to make the task of acquiring the parts necessary for construction of the projects described in this book easier, the author has included part numbers and in some cases hyperlinks to various vendors. These are strictly for the convenience of the reader.
- The URLs provided in this book were valid at the date of publication; however, Web links may change over time.
- The author does not endorse the use of any particular vendor, or discourage the use of any vendor not mentioned in this book.
- Where a vendor is listed for a part, the author—in preparation for this book—used that vendor to acquire the part.
- The author has received no compensation from any organization or company for any aspect of this book, other than the publisher.

- Readers acknowledge that working with electrical devices carries potential harm. The circuits described in this book utilize voltages sufficient to result in injury or death. The author has included cautionary statements to this effect in multiple chapters of this book.
- Readers' attention is called to the discussion of the dangers of electrical shock contained in Chapter 6. This chapter is intended as an overview of the subject. It is not intended to comprehensively deal with the potential results of electrical shock or first aid procedures. Readers are encouraged to explore any of the many resources in this area.

The author would assume that such clarifying statements would (or at least should) be contained in any book dealing with the construction of electronics projects. In an event, we have now covered them. The intent here is to put the "fine print" up front and in a normal font size. Having dealt with the preliminary steps, let's get started!

Chapter 1

An Overview of Vacuum Tube Audio Applications

The phrase "high technology" is perhaps one of the more overused descriptions in our technical vocabulary. It is a phrase generally reserved for the discussion of integrated circuits, fiber optics, satellite systems, computers, and handheld portable devices of many varieties. Very few people would associate high technology with vacuum tubes—except audio enthusiasts. Variously described as the "tube sound," amplifiers built around vacuum tubes remain in demand for demanding consumers.

A number of projects are included in this book. Several of those projects—in finished form—are shown in Figure 1.1.

The Evolution of Analog Audio

The use of solid-state technology in all manner of consumer audio devices has made possible the explosion of audio sources and options for consumers—at very attractive prices. It is difficult to imagine a world without personal entertainment devices—although it is sometimes tempting to do so. Whether the personalization of entertainment is a good thing or a not-so-good thing could be debated, probably at some length. It's all academic, of course, since it is here.

Acknowledging up front that this is a book about technology, it is fair to point out that to the end user, audio—perhaps more than any other entertainment medium—is about preferences and real-life experiences. Audio has certain fundamental reference points—loudness, frequency response, noise, distortion, and so forth. It has another dimension as well, and that dimension is perception. With audio, the artist and/or producer has a wide and varied pallet with which to paint. There are few absolutes when it comes to audio perception. With video, on the other hand, absolutes abound. Viewers know that the grass should be green and the sky should be blue and people should look like ... people. Audio has the capacity for texture and subtlety, which frankly makes it more interesting.

FIGURE 1.1   Some of the vacuum tube audio amplifier projects detailed in this book—in finished form.

Audio is, of course, more than music. However, music makes up a large part of what we consider audio and what consumers use the technologies of audio for. The social impact of audio (music) should not be underestimated. Music provides reference points for our lives. Most everybody can relate to hearing a song play and reflecting back to a particular event in their life—sometimes from the very distant past. This social aspect was probably more profound in the era of the 1950s through the 1970s when most listening to music was a group event focusing on a limited number of radio stations. When a new album was released by a given performer, most everyone in a particular age group heard it and reacted to it. For better or worse, this gave generations of listeners various reference points to which they can still relate. The revolution in personal entertainment devices has, to a large degree, diluted this group experience. Whether this is important to anybody remains to be seen.

It is easy to argue that consumer audio has been in a long march toward the lowest common denominator, focused on cost and size more than performance. Others may wish to debate that, but regardless, the manufacturers in this space have been giving consumers what they want. And it's hard to argue with that. Still, for consumers who are looking for more than just convenience from their audio system, options are—thankfully—still readily available.

The author, like some percentage of readers of this book (perhaps a large percentage), grew up in the 1960s. With an interest in electronics, that meant also an interest in vacuum tubes. By 1970, consumer electronics manufacturers had moved in large measure to transistorized amplifiers, tuners, portable radios, and so on. Intrigued by this new technology, few of us realized what we were giving up by discarding vacuum tube equipment in favor of new solid-state hardware. But because personal storage space is never unlimited, the old stuff went away in favor of new stuff (which went away later). If you are reading this book (and there is a good chance that you are), you, like the author, have rediscovered what we all thought was out of date three or four decades ago.

The author can remember in the early 1970s literally running away from vacuum tube circuits to embrace transistor-based circuits. Transistors were—of course—better, smaller, cheaper, newer. And, most importantly, solid-state circuits always provided improved performance over their tube counterparts. Or so we all thought (or at least a lot of us thought). Looking back now with the benefit of history, it is clear that solid-state devices did some things very well and tubes did other things very well. Today, there is room for both in any entertainment center.

In preparation for this book, the author began collecting various types of vintage tube hardware from eBay and other sources with the intent of refurbishing it. These projects served as a reminder that: 1) this stuff is fun to work with; 2) circuits based on vacuum tubes are interesting; 3) circuits based on vacuum tubes are understandable (in contrast to many products today, which are really understood by only a very small number of people); and 4) vacuum tubes are actually quite reliable. The last point deserves some elaboration. The author, in his refurbishing projects, has found (to some surprise) that a lot of hardware built 50 years ago will actually still light up and do something useful (if the old electrolytics do not smoke first). The surprising find was that tubes, when properly cared for, actually *were* reliable. And they still are.

Technology Waves

It is useful to review the progression of consumer audio devices over the past few decades. They have tended to come in a series of waves. Various benchmarks or inflection points for consumer audio technologies can be identified; however, as a first-order approximation, the following general divisions seem to cover most of the bases.

Pre-1950

Characterized by physically large systems with limited features, pre-1950s audio equipment was usually nothing to write home about—or remember for that matter. These sets were, by and large, big pieces of furniture designed for one or two functions—radio and/or records. One bright spot in the stock receiver, however, was the typical frequency response of the AM radio circuits. During the 1950s as the number of AM radio stations grew rapidly, set designers had to adjust their filtering schemes to accommodate additional interference from nearby (and at night distant) stations. One common approach was to limit the bandwidth of the received signal. This limited interference, but, of course, also limited frequency response. The growth of AM radio was a classic case where success was not necessarily a good thing, at least from the standpoint of sonic performance. The pre-1950s receivers were invariably intended for fixed operation, some with the capability to add an external long-wire antenna for reception of shortwave broadcasts.

1950s Audio

The evolution in consumer electronics continued with improved receivers (frequency response notwithstanding) and higher-quality turntables. While the "45" record came on strong in the 1950s, the LP had also firmly established itself as the medium for high-fidelity listening at home. The large furniture-piece sets began to give way to smaller single-function devices. The extra space in the living room was, of course, quickly consumed by television receivers. (We won't discuss their audio performance here.) Portable radios also appeared, using tubes at first, and powered by large dry-cell batteries (not rechargeable). Radios also started appearing in automobiles in large numbers, typically using innovative (if not elegant) methods of generating the necessary operating voltages for tubes from a 12 V DC power source (enter the "vibrator" device that chopped the direct-current source to simulate an alternating-current source, which was applied to a step-up transformer).

High-end audio equipment began appearing, marketed to an emerging discriminating audience. Some of these systems were very good—very, very good. With 3 dB frequency response points of less than 10 Hz and greater than 100 kHz, these amplifiers set the pace for high-fidelity systems that followed. This development is even more impressive when considering that virtually no source material existed that would fully take advantage of the capabilities of the amplifier—certainly not AM radio or even the developing FM radio, and certainly not the vinyl records then available.

1960s Audio

Behold the transistor radio. Small enough to fit in a shirt pocket, the capabilities of this miniature marvel were often described in terms of the number of transistors used. One would assume that a seven-transistor radio was better than a five-transistor radio. With their two-inch speakers, it was hard to tell the difference anyway. But, this new device gave consumers portability, and they liked it. Despite the inroads made by transistors, vacuum tubes still reigned supreme. Audio systems with high-quality turntables as the input source moved into mainstream use and ushered in a (here it comes) "golden era" of audio.

On the receiver side, reduction in size seemed to be the main trend, exemplified by the five-tube table-top AM radio shown in schematic form in Figure 1.2. The high-end models included a clock and "wake to music" alarm. FM receivers began to appear in large numbers, driven by the high-quality audio (and even stereo) programming available. The growing number of counterculture rock music stations didn't hurt either.

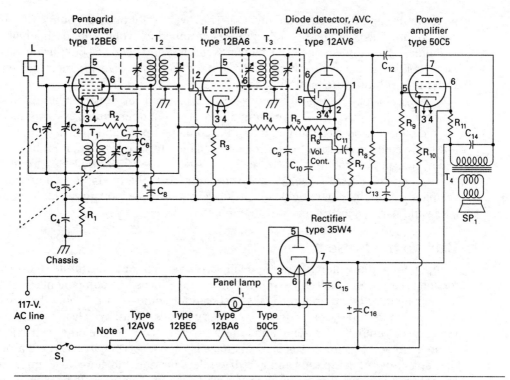

FIGURE 1.2   Schematic diagram of the classic five-tube table-top AM radio. (*From* [1].)

Reel-to-reel tape recorders also appeared at reasonable prices for consumers, with a limited selection of music available on reels of various sizes. Readers will recall the three common speeds (1-7/8, 3-3/4, and 7-1/2 inches per second) and the various reel sizes (3-1/2, 5, and 7 inches for consumer products).

Console Audio

The furniture radio set of 1950 came back in the 1960s as an entertainment center that included, depending on the model, an AM receiver, FM receiver, turntable, and perhaps television set. One could argue this was the last stand for audio furniture in the home. While the focus of these systems seemed to be mostly features and convenience rather than overall performance, they were the focal point of countless living rooms for many years. They also served to advance the concept of high-quality audio entertainment for consumers who were enamored with the shirt-pocket transistorized AM radio but recognized that some things are worth sitting down to listen to.

Component Audio

The 1970s were all about component audio. Designers took the console audio systems of the 1960s and broke them into discrete devices, reasoning they could enhance the performance and features in the process. They were right on both counts. Consumers loved them. The component audio system continued to evolve and reach a high level of sophistication and sonic performance. Systems could be found built using vacuum tubes, solid-state devices, or both. Some of the most innovative and memorable audio systems ever built were built in this era.

The genius of component audio was that it allowed consumers to build a system over time into exactly what they wanted it to be. Component audio systems also offered considerable flexibility in that units could be mixed and matched to yield just the system envisioned by the consumer. The interface/connector problems that bedevil consumers today were not really an issue in the 1970s and 1980s, as nearly every input and output (other than the speakers) used the trusty RCA connector. Couple this simplicity with look-alike styling that produced an attractive tower of audio and happy consumers were guaranteed.

Back to the Present

Component audio systems were the mainstay of consumer audio for decades—and for good reason. But like every trend in home appliances, the component system has been challenged by other approaches to consumer audio—most notably the personal audio player and the wide variety of accessory devices that have clustered around the player to embellish, extend, and otherwise enhance it. At the other extreme, the component architecture has been challenged by home server systems that harness the capabilities of computers and wired/wireless networks to store huge amounts of content and move it more or less seamlessly around the house to be consumed privately or collectively.

Along with the move toward personal audio (and video) players has come the notion that small is good—at least as it relates to audio entertainment devices. (The size of flatscreen video displays is another matter entirely—limited only by wall size and available funds.) The bookshelf (or even floor-standing) speakers that dominated living spaces in the 1980s and 1990s have been replaced in many homes by small speakers utilizing a common subwoofer. For a number of consumers, the performance is good enough and the prices are certainly attractive.

Having acknowledged there are trends, of course, does not mean everyone needs to follow them.

Tube vs. Solid State

The corner piece of the component audio system is—inevitably—the power amplifier. As transistors replaced tubes and integrated circuits replaced transistors and surface-mounted devices replaced integrated circuits, the measurable performance of audio systems has steadily improved, sometimes dramatically so. The benefits of solid-state technologies in low-level audio circuits are well known, beginning with noise and distortion performance. Similar attributes apply to radio frequency (RF) circuits. In the case of disc players (CD and DVD) it's all about data, and what we generally consider audio plays a relatively minor role at the end of a long chain of logic gates.

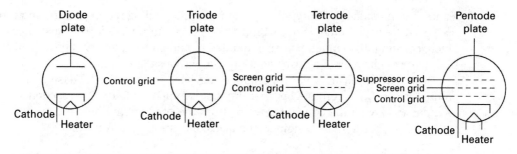

FIGURE 1.3 The primary types of vacuum tubes. Variations also exist where combinations of these basic elements are enclosed within the glass envelope.

As for the power amplifier, however, the choice between solid-state and vacuum tubes is not so clear-cut.

While it is certainly true that solid-state preamplifiers (preamp) and power amplifiers solved a host of shortcomings of their predecessor vacuum tube amplifiers, a certain sonic quality was lost in the process. Often described in nonscientific terms, the "warmth" of the tube sound nonetheless exists and has attracted a loyal following. Vacuum tube–based audio equipment remains in demand and is likely to remain so for a long time.

Various explanations have been offered over the years to define the tube sound and how it differs from audio produced using solid-state devices. This comparison is made more difficult by the differing amplifier architectures that have been used to construct transistor-based power amplifiers in order to improve efficiency and/or measured performance. It is arguable, however, that identifying the characteristics that define the differences is not really all that important. It is probably sufficient to simply acknowledge that there are differences and accept them. As noted earlier in this chapter, audio is all about how humans react to it. How a selection of music is heard (perceived) is—in the end—more important than the documented transfer characteristics of the active devices contained in the amplifier.

Viewed objectively, it is evident that some very good vacuum tube amplifiers were offered to consumers, and some very bad ones were offered as well. The tradeoff between price point and performance is neither new nor limited to electronic devices. By the same token, there have been some very good solid-state amplifiers offered to consumers, and some very bad ones too.

Vacuum tubes (known in earlier times as "receiving tubes" or alternately as "valves") include a wide range of devices, each for a specific class of applications. Devices include diodes, triodes, tetrodes, and pentodes, as illustrated in Figure 1.3. These devices hold an important position in high-fidelity audio.

Engineering Tradeoffs

Designing a consumer product is almost always an exercise in tradeoffs. The variables include (but are certainly not limited to) complexity, component count, bill of materials, feature set, manufacturability, power requirements, cooling requirements, and time-to-market. A "perfect" audio device would likely be a commercial flop because it cost too much, consumed too much power, generated too much heat, and was never finished. Well, that may be a bit of an exaggeration, but perhaps not by much.

Design tradeoffs are a part of engineering. The first step in the design of any product is to clearly define what that product is supposed to do. The second step is to understand what the consumer wants from the product and how much they are willing to pay for it. Other engineering decisions branch out from there.

Basic System Choices

Focusing on vacuum tube audio amplifiers in general, and this book in particular, before setting out to build an amplifier, it is necessary to answer some basic questions:

- *What is the intended application?* Options include: 1) turntable preamp, 2) microphone preamp, 3) line-level preamp, 4) equalizer (tone control) preamp, or 5) power amplifier (and, of course, a power supply needed to make options 1–5 work). As a practical matter, any system will likely include some combination of these functions, and perhaps other functions as well. For the purpose of this book we will plan on all six functions listed here.
- *What are the desired active devices?* Options include: 1) solid-state, 2) vacuum tube, or 3) a combination of each. For a consumer product, this choice is fundamental and driven by many factors—some technical, some not. For the purpose of this book, however, it is clear the desired active devices will be vacuum tubes or a combination of tubes and solid-state devices. For a power amplifier, the required output level is a fundamental consideration that determines the overall architecture of the system (e.g., single-ended, push-pull, parallel, etc.).
- *What is the intended form factor?* Options include: 1) stand-alone device, 2) component system, 3) integrated with another device (such as a turntable or speaker), or 4) something else. The form factor dictates a major cost of the project—namely, the physical enclosure (case). For this book we will assume a component system built using off-the-shelf sheet metal components with limited custom cutting as needed.
- *What is the preferred construction method?* Options include: 1) printed wiring board (PWB), 2) hand-wired, or 3) a combination of both techniques. Any of the three approaches are practical for vacuum tube designs. While the PWB method results in simpler construction (once the PWB has been designed and produced) and a neater appearance, the heat generated by vacuum tubes must be considered. As such, a hybrid approach where the tubes are mounted on the metal chassis and tied by interconnecting wires to the PWB may be preferred. A completely hand-wired chassis is also an option—perhaps the sentimental favorite given the history of the vacuum tube. For experimentation or a one-off project, it's hard to beat a hand-wired chassis for low cost, simplicity, and long-term performance.
- *Will the design be limited to off-the-shelf components?* In an effort to extract the last measure of performance from a tube amplifier, some audiophiles choose to build custom components, such as winding the output transformer to certain

specifications. While the benefits of this approach can be significant for the experienced builder, for the purpose of this book the scope of the projects will be limited to off-the-shelf components. This approach makes certain assumptions about the level of detail that most readers are interested in for most projects most of the time.

- *What is the maximum estimated cost for the finished unit?* With any project or product, it is rare to encounter a situation where money is not a consideration. For the purpose of this book, unlimited funds will not be assumed. It is also important to point out that the estimated cost for most any project is often exceeded well before the project has been completed.
- *How much time is available to build the unit?* Like unlimited funds, unlimited time is usually in short supply as well. For this book it is assumed that while building an audio amplifier should be an enjoyable project, it should not consume the builder's life.
- *Do I have the technical ability to do this?* Yes, of course you do!

Continuing Development of Vacuum Tubes

In the realm of vacuum tubes, there is a natural division between two fundamental classes of devices: receiving tubes and power tubes. Power vacuum tubes cover a wide range of devices, many exotic, and are still used in countless applications. Due to improvements in solid-state devices, power vacuum tube development has focused on high powers and high frequencies where their unique advantages can be exploited. The power levels possible with vacuum devices are truly astounding—many hundreds of kilowatts power output from a single device is not uncommon at ultra-high frequencies (UHF). No solid-state device can match this level of performance. This being the case, tube development continues as engineers push the limits of power output, maximum operating frequency, efficiency, and reliability.

The primary frontier in power vacuum tube development today is materials technology. New and improved devices depend on new and improved materials. This is an exciting area of applied science where evolutionary progress continues to be made.

Receiving tubes do not enjoy the same developmental effort on new devices, but producers are still exploring ways of improving classic models and optimizing production techniques. The continued demand for receiving tubes drives this work. Numerous retail sources exist for receiving tubes, both new and vintage stock.

Apart from continuing refinements in manufacturing methods, the last major technology advancement for the receiving tube was probably the "integrated vacuum tube." This class of device, known by various trade names (e.g., "Compactron"), took the elements of various basic tube types and combined them to form a larger device that performed several functions. One example of this class of device is shown in Figure 1.4.

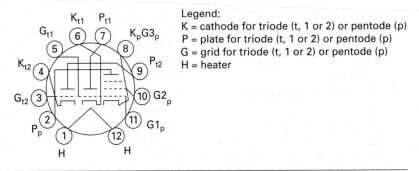

FIGURE 1.4 Example "integrated vacuum tube" containing two triodes and a pentode within a single glass envelope. This particular device (14BL11) was used in television receiver applications. (*From* [1]).

Standardization

The need for interchangeability of vacuum tubes in the 1930s and especially the 1940s for military applications drove product standardization and helped make tubes synonymous with electronics. Without a standardized scheme for device labeling, performance, and interconnection, the great advances in electronics made during this period would have been difficult—perhaps impossible.

Product standardization is important for the advancement of any industry—particularly the electronics industry where a large number of individual components is required to construct any single product. Standardization leads to a healthy commercial environment where multiple vendors work to develop new techniques and technologies that advance the science of component design and fabrication. In addition, having multiple vendors for a given device usually results in lower prices to the user.

Nomenclature

If a vacuum tube users group were to sit down and define a numbering/identification system for tubes today, it probably would come up with something quite different from what we actually have. One could imagine a nomenclature that would convey a great deal of information about the device itself. While imperfect, vacuum tube nomenclature is nonetheless stable and predictable, and given the number of devices typically used for audio applications, is quite manageable.

A user can typically assume that the same device type from any of several manufacturers will provide similar nominal performance. As with any product, of course, manufacturers seek to differentiate their offerings from competitors through various attributes, such as higher performance, longer life, and so on. For critical applications it may be necessary to use a particular brand of device to achieve the performance desired. By design, the applications contained in this book will not fall into the "critical performance" category.

It is fair to point out that some users have preferences for particular device brands, and while it may be hard to characterize empirically what those differences are, such preferences are nonetheless valid. Audio, after all, is all about perception of the reproduced sound. Personal preference and past experience can be more significant than any specifications sheet.

Fundamental Electrical Principles Reviewed

Before moving into vacuum tube theory, circuit design, and construction projects, it is worthwhile to review some of the fundamental electrical principles that all readers learned many years ago (perhaps decades ago). A short refresher is probably a good idea.

The Atom

The atomic theory of matter specifies that each of the many chemical elements is composed of unique and identifiable particles called *atoms* [2]. In ancient times only ten were known in their pure, uncombined form; these were carbon, sulfur, copper, antimony, iron, tin, gold, silver, mercury, and lead. Of the several hundred now identified, fewer than 50 are found in an uncombined, or chemically free, form on earth.

Each atom consists of a compact *nucleus* of positively and negatively charged particles (*protons* and *electrons*, respectively). Additional electrons travel in well-defined orbits around the nucleus. The electron orbits are grouped in regions called *shells*, and the number of electrons in each orbit increases with the increase in orbit diameter in accordance with quantum-theory laws of physics. The diameter of the outer orbiting path of electrons in an atom is in the order of one-millionth (10^{-6}) millimeter, and the nucleus, one-millionth of that. These typical figures emphasize the minute size of the atom.

Magnetic Effects

The nucleus and the free electrons of an iron atom are shown in the schematic diagram in Figure 1.5 [2]. Note that the electrons are spinning in different directions. This rotation creates a magnetic field surrounding each electron. If the number of electrons with positive spins is equal to the number with negative spins, then the net field is zero and the atom exhibits no magnetic field.

In the diagram, although the electrons in the first, second, and fourth shells balance each other, in the third shell five electrons have clockwise positive spins, and one a counterclockwise negative spin, which gives the iron atom in this particular electron configuration a cumulative *magnetic effect*.

The parallel alignment of electrons spins over regions, known as *domains*, containing a large number of atoms. When a magnetic material is in a demagnetized state, the direction of magnetization in the domain is in a random order. Magnetization by an external field takes place by a change or displacement in the isolation of the domains, with the result that a large number of the atoms are aligned with their charged electrons in parallel.

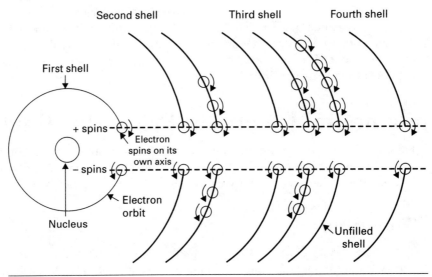

FIGURE 1.5   Schematic of the iron (Fe) atom. (*After* [2].)

Conductors and Insulators

In some elements, such as copper, the electrons in the outer shells of the atom are so weakly bound to the nucleus that they can be released by a small electrical force, or voltage [2]. A voltage applied between two points on a length of a metallic conductor produces the flow of an electric current, and an electric field is established around the conductor. The *conductivity* is a constant for each metal that is unaffected by the current through or the intensity of any external electric field.

In some nonmetallic materials, the free electrons are so tightly bound by forces in the atom that, upon the application of an external voltage, they will not separate from their atom except by an electrical force strong enough to destroy the insulating properties of the material. However, the charges will realign within the structure of their atom. This condition occurs in the insulating material (*dielectric*) of a capacitor when a voltage is applied to the two conductors encasing the dielectric.

Semiconductors are electronic conducting materials wherein the conductivity is dependent primarily upon impurities in the material. In addition to negative mobile charges of electrons, positive mobile charges are present. These positive charges are called *holes* because each exists as an absence of electrons. Holes (+) and electrons (–), because they are oppositely charged, move in opposite directions in an electric field. The conductivity of semiconductors is highly sensitive to, and increases with, temperature.

Direct Current (DC)

Direct current is defined as a unidirectional current in which there are no significant changes in the current flow [2]. In practice, the term frequently is used to identify a voltage source, in which case variations in the load can result in fluctuations in the current but not in the direction.

Direct current was used in the first systems built to distribute electricity for household and industrial power. For safety reasons, and the voltage requirements of lamps and motors, distribution was at the low nominal voltage of 110 V. The losses in distribution circuits at this voltage seriously restricted the length of transmission lines and the size of the areas that could be covered. Consequently, only a relatively small area could be served by a single generating plant. It was not until the development of alternating-current systems and the voltage transformer that it was feasible to transport high levels of power at relatively low current over long distances for subsequent low-voltage distribution to consumers.

Alternating Current (AC)

Alternating current is defined as a current that reverses direction at a periodic rate [2]. The average value of alternating current over a period of one cycle is equal to zero. The effective value of an alternating current in the supply of energy is measured in terms of the *root mean square* (rms) value. The rms is the square root of the square of all the values, positive and negative, during a complete cycle, usually a sine wave. Because rms values cannot be added directly, it is necessary to perform an rms addition as follows:

$$V_{\text{rms total}} = \sqrt{V_{\text{rms1}}^2 + V_{\text{rms2}}^2 + \ldots V_{\text{rms n}}^2} V$$

As in the definition of direct current, in practice the term frequently is used to identify a voltage source.

The level of a sine-wave alternating current or voltage can be specified by two other methods of measurement in addition to rms. These are *average* and *peak*. A sine-wave signal and the rms and average levels are shown in Figure 1.6. The levels of complex, symmetrical AC signals are specified as the peak level from the axis, as shown in the figure.

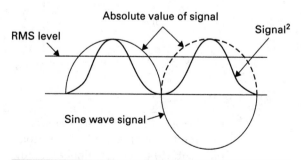

FIGURE 1.6   Root mean square (rms) measurements. The relationship of rms and average values is shown. (*After* [2].)

Electronic Circuits

Electronic circuits are composed of elements such as resistors, capacitors, inductors, and voltage and current sources, all of which may be interconnected to permit the flow of electric currents [2]. An *element* is the smallest component into which circuits can be subdivided. The points on a circuit element where they are connected in a circuit are called *terminals*.

Elements can have two or more terminals, as shown in Figure 1.7. The resistor, capacitor, inductor, and diode shown in Figure 1.7a are two-terminal elements; the transistor in Figure 1.7b is a three-terminal element; and the transformer in Figure 1.7c is a four-terminal element.

Circuit elements and components also are classified as to their function in a circuit. An element is considered *passive* if it absorbs energy and *active* if it increases the level of energy in a signal. An element that receives energy from either a passive or active element is called a *load*. In addition, either passive or active elements, or components, can serve as loads.

The basic relationship of current and voltage in a two-terminal circuit where the voltage is constant and there is only one source of voltage is given in Ohm's law. This states that the voltage E between the terminals of a conductor varies in accordance with the current I. The ratio of voltage, current, and resistance R is expressed in Ohm's law:

$$E = I \times R$$

Using Ohm's law, the calculation for power in watts can be developed from $P = E \times I$ as follows:

$$P = \frac{E^2}{R} \text{ and } P = I^2 \times R$$

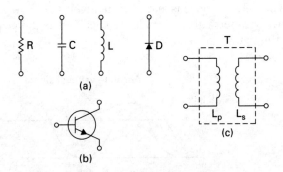

FIGURE 1.7   Schematic examples of circuit elements: (*a*) two-terminal element, (*b*) three-terminal element, (*c*) four-terminal element. (*After* [2].)

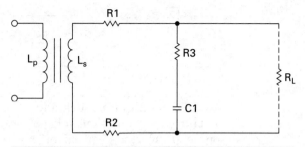

FIGURE 1.8   Circuit configuration composed of several elements and branches and a closed loop (R1, R3, C1, R2, and L$_s$). (*After* [2].)

A circuit, consisting of a number of elements or components, usually amplifies or modifies a signal before delivering it to a load. The terminal to which a signal is applied is an *input port,* or *driving port.* The pair or group of terminals that delivers a signal to a load is the *output port.* An element or portion of a circuit between two terminals is a *branch.* The circuit shown in Figure 1.8 is made up of several elements and branches. R1 is a branch, and R3 and C1 make up a two-element branch. The secondary winding of transformer, a voltage source, and R2 also constitute a branch. The point at which three or more branches join together is a *node.* A series connection of elements or branches, called a *path,* in which the end is connected back to the start, is a *closed loop.*

Circuit Analysis

Relatively complex configurations of linear circuit elements (e.g., where the signal gain or loss is constant over the signal amplitude range) can be analyzed by simplifying them into the equivalent circuits [2]. After restructuring a circuit into an equivalent form, the current and voltage characteristics at various nodes can be calculated using network-analysis theorems, including Kirchhoff's current and voltage laws, Thevenin's theorem, and Norton's theorem.

- **Kirchhoff's current law** (KCL)   The algebraic sum of the instantaneous currents entering a node (a common terminal of three or more branches) is zero. In other words, the currents from two branches entering a node add algebraically to the current, leaving the node in a third branch.
- **Kirchhoff's voltage law** (KVL)   The algebraic sum of instantaneous voltages around a closed loop is zero.
- **Thevenin's theorem**   The behavior of a circuit at its terminals can be simulated by replacement with a voltage E from a DC source in series with an impedance Z (see Figure 1.9*a*).
- **Norton's theorem**   The behavior of a circuit at its terminals can be simulated by replacement with a DC source I in parallel with an impedance Z (see Figure 1.9*b*).

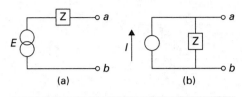

FIGURE 1.9   Equivalent circuits:
(*a*) Thevenin's equivalent voltage
source, (*b*) Norton's equivalent
current source. (*After* [3].)

Static Electricity

The phenomenon of static electricity and related potential differences concerns
configurations of conductors and insulators where no current flows and all electrical
forces are unchanging; hence the term *static* [2]. Nevertheless, static forces are present
because of the number of excess electrons or protons in an object. A static charge can
be induced by applying voltage to an object. A flow of current to or from the object
can result from either a breakdown of the surrounding nonconducting material or by
the connection of a conductor to the object.

Two basic laws regarding electrons and protons are

- Like charges exert a repelling force on each other; electrons repel other electrons
  and protons repel other protons.
- Opposite charges attract each other; electrons and protons are attracted to
  each other.

Therefore, if two objects each contain exactly as many electrons as protons in each
atom, there is no electrostatic force between the two. On the other hand, if one object
is charged with an excess of protons (deficiency of electrons) and the other an excess
of electrons, there will be a relatively weak attraction that diminishes rapidly with
distance. An attraction also will occur between a neutral and a charged object.

Another fundamental law governing static electricity, developed by Faraday, is
that all of the charge of any conductor not carrying a current lies in the surface of the
conductor. Thus, any electric fields external to a completely enclosed metal box will
not penetrate beyond the surface. Conversely, fields within the box will not exert any
force on objects outside the box. The box need not be a solid surface; a conduction cage
or grid will suffice. This type of isolation frequently is referred to as a *Faraday shield*.

Magnetism

The elemental magnetic particle is the spinning electron [2]. In magnetic materials,
such as iron, cobalt, and nickel, the electrons in the third shell of the atom are the
source of magnetic properties. If the spins are arranged in parallel, the atom and its
associated domains or clusters of the material will exhibit a magnetic field. The magnetic

field of a magnetized bar has lines of magnetic force that extend between the ends, one called the north pole and the other the south pole, as shown in Figure 1.10a. The lines of force of a magnetic field are called *magnetic flux* lines.

Electromagnetism

A current flowing in a conductor produces a magnetic field surrounding the wire as shown in Figure 1.11a [2]. In a coil or solenoid, the direction of the magnetic field relative to the electron flow (– to +) is shown in Figure 1.11b. The attraction and repulsion between two iron-core electromagnetic solenoids driven by direct currents is similar to that of two permanent magnets described previously.

The process of magnetizing and demagnetizing an iron-core solenoid using a current applied to a surrounding coil can be shown graphically as a plot of the magnetizing field strength and the resultant magnetization of the material, called a *hysteresis loop* (Figure 1.12). At the point where the field is reduced to zero, a small amount of magnetization, called *remnance,* remains.

Magnetic Shielding

In effect, the shielding of components and circuits from magnetic fields is accomplished by introducing a magnetic short circuit in the path between the field source and the area to be protected [2]. The flux from a field can be redirected to flow in a partition or shield of magnetic material, rather than in the normal distribution pattern between north and south poles. The effectiveness of shielding depends primarily upon the thickness of the shield, the material, and the strength of the interfering field.

Some alloys are more effective than iron. However, many are less effective at high flux levels. Two or more layers of shielding, insulated to prevent circulating currents from magnetization of the shielding, may be used in low-level audio, video, and data applications.

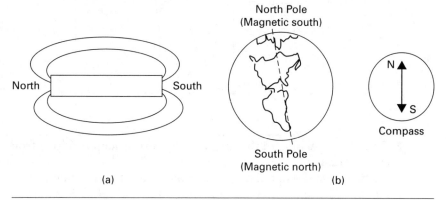

FIGURE 1.10   The properties of magnetism: (*a*) lines of force surrounding a bar magnet, (*b*) relation of compass poles to the earth's magnetic field. (*After* [2].)

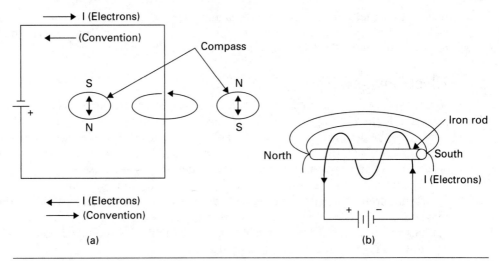

FIGURE 1.11   Magnetic field surrounding a current-carrying conductor:
(*a*) Compass at right indicates the polarity and direction of a magnetic field
circling a conductor carrying direct current. *I* indicates the direction of electron
flow. Note: The convention for flow of electricity is from + to –, the reverse of
the actual flow. (*b*) Direction of magnetic field for a coil or solenoid. (*After* [2].)

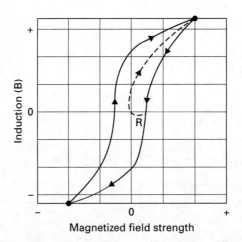

FIGURE 1.12   Graph of the magnetic hysteresis loop resulting from
magnetization and demagnetization of iron. The dashed line is a plot of the
induction from the initial magnetization. The solid line shows a reversal of
the field and a return to the initial magnetization value. *R* is the remaining
magnetization (remnance) when the field is reduced to zero. (*After* [2].)

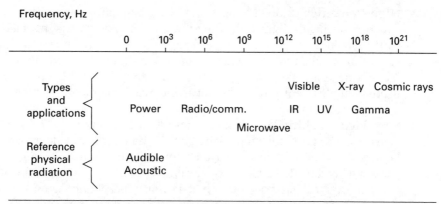

FIGURE 1.13 The electromagnetic spectrum. (*After* [3].)

Electromagnetic-Radiation Spectrum

The usable spectrum of electromagnetic-radiation frequencies extends over a range from below 100 Hz for power distribution to 10^{20} for the shortest X-rays [2]. Services using various frequency bands in the spectrum are shown in Figure 1.13. The lower frequencies are used primarily for terrestrial broadcasting and communications. The higher frequencies include visible and near-visible infrared and ultraviolet light, and X-rays.

The electromagnetic spectrum can be roughly divided into the following general categories:[1]

- **Low-end spectrum frequencies** (1 to 1000 Hz) Electric power is transmitted by wire but not by radiation at 50 and 60 Hz, and in some limited areas, at 25 Hz. Aircraft use 400-Hz power in order to reduce the weight of iron in generators and transformers. The restricted bandwidth that would be available for communication channels is generally inadequate for voice or data transmission, although some use has been made of communication over power distribution circuits using modulated carrier frequencies. The sound-transmission frequencies noted in Figure 1.13 are acoustic rather than electromagnetic.
- **Low-end radio frequencies** (1000 to 100 kHz) These low frequencies are used for very long-distance radio-telegraphic communication where extreme reliability is required and where high-power long antennas can be erected.
- **Medium-frequency radio** (100 kHz to 2 MHz) The low-frequency portion of the band is used for around-the-clock communication services over moderately long distances and where adequate power is available to overcome the high level of atmospheric noise. The upper portion is used for AM radio, although the strong and quite variable sky wave occurring during the night results in substandard quality and severe fading at times. Other uses include fixed and mobile service and amateur radio communication.

---

[1] Note that specific spectrum division and classification terms have been developed and are in use worldwide. For our purposes here, however, these convenient general groupings will suffice.

- **High-frequency radio** (2 to 30 MHz)   This band provides reliable medium-range coverage during daylight and, when the transmission path is in total darkness, worldwide long-distance service, although the reliability and signal quality of the latter is dependent to a large degree upon ionospheric conditions and related long-term variations in sunspot activity affecting sky-wave propagation. The primary applications include broadcasting, fixed and mobile services, telemetry, and amateur transmissions.
- **Very high and ultra-high frequencies** (30 MHz to 3 GHz)   VHF and UHF bands, because of the greater channel bandwidth possible, can provide transmission of large amounts of information. Furthermore, the shorter wavelengths permit the use of highly directional parabolic or multielement antennas. Reliable long-distance communication is provided using high-power tropospheric scatter techniques. The multitude of uses include television, fixed and mobile communication services, amateur radio, radio astronomy, satellite communication, telemetry, and radar.
- **Microwaves** (3 to 300 GHz)   At these frequencies, many transmission characteristics are similar to those used for shorter optical waves, which limit the distances covered to line of sight. Typical uses include microwave relay, satellite, radar, and wide-band information services.
- **Infrared, visible, and ultraviolet light**   The portion of the spectrum visible to the eye covers the gamut of transmitted colors ranging from red, through yellow, green, and blue. It is bracketed by infrared on the low-frequency side and ultraviolet (UV) on the high side. Infrared signals are used in a variety of consumer and industrial equipments for remote controls and sensor circuits.
- **X-rays**   Medical and biological examination techniques and industrial and security inspection systems are the best-known applications of X-rays. X-rays in the higher-frequency range are classified as *hard* X-rays or *gamma* rays. Exposure to X-rays for long periods can result in serious irreversible damage to living cells or organisms.

Audio Spectrum

For the purposes of this book, we will focus on the audio spectrum, which is generally accepted to range from 20 Hz to 20 kHz. It is important to note that advances in analog and digital technologies have pushed the lower limit of the audio spectrum to near DC—not that anyone can hear it (although listeners can *feel* it). Such advances have also extended the high end to well above 20 kHz—again, not that anyone can hear it. For digital systems sampled at 48 kHz (a common reference for professional applications), the theoretical upper boundary is 24 kHz due to the Nyquist limit[2].

In any chain of devices, the overall performance of the system is limited by the weakest link. In the case of vacuum tube audio amplifiers, the weakest link insofar

---

[2] The Nyquist law for digital coding dictates that the sample rate must be at least twice the cutoff frequency of the signal of interest to avoid spurious patterns (aliasing) generated by the interaction between the sampling signal and the higher signal frequencies.

as bandwidth is concerned is usually the output transformer. The practical realities of transformer design and construction make it difficult to reproduce very low frequencies or very high frequencies. Having said that, a number of high-performance transformers have been developed that will faithfully reproduce waveforms from 10 Hz (or less) to more than 50 kHz (in a few cases well above 50 kHz).

Decibel Measurement

Audio signals span a wide range of levels [4]. The sound pressure of a rock-and-roll band is about 1 million times that of rustling leaves. This range is too wide to be conveniently accommodated on a linear scale. The decibel is a logarithmic unit that compresses this wide range down to a more easily handled range. Order-of-magnitude (factor-of-10) changes result in equal increments on a decibel scale. Furthermore, the human ear perceives changes in amplitude on a logarithmic basis, making measurements with the decibel scale reflect audibility more accurately.

A decibel may be defined as the logarithmic ratio of two power measurements or as the logarithmic ratio of two voltage measurements. The following equations define the decibel for voltage (E) and power (P):

$$dB = 20 \log \frac{E_1}{E_2}$$

$$dB = 10 \log \frac{P_1}{P_2}$$

There is no difference between decibel values from power measurements and from voltage measurements if the impedances are equal. In both equations, the denominator variable is usually a stated reference. Whether the decibel value is computed from the power-based equation or from the voltage-based equation, the result is the same.

A doubling of voltage will yield a value of 6.02 dB, and a doubling of power will yield 3.01 dB. This is true because doubling voltage results in a factor-of-4 increase in power. Table 1.1 shows the decibel values for some common voltage and power ratios.

Audio engineers often express the decibel value of a signal relative to some standard reference, rather than another signal. The reference for decibel measurements may be predefined as a power level, as in dBm (decibels above 1 mW), or it may be a voltage reference. When measuring dBm or any power-based decibel value, the reference impedance must be specified or understood.

It is often desirable to specify levels in terms of a reference transmission level somewhere in the system under test. These measurements are designated dBr, where the reference point or level is separately conveyed.

TABLE 1.1   Common Decibel Values and
Conversion Ratios

dB Value	Voltage Ratio	Power Ratio
–40	0.01	0.0001
–20	0.1	0.01
–10	0.3163	0.1
–6	0.501	0.251
–3	0.707	0.501
–2	0.794	0.631
–1	0.891	0.794
0	1	1
+1	1.122	1.259
+2	1.259	1.586
+3	1.412	1.995
+6	1.995	3.981
+10	3.162	10
+20	10	100
+40	100	10,000

Dimensions of Hearing

The perception of sound is a complex process involving many variables [5]. *Loudness* is the term used to describe the magnitude of an auditory sensation. It is primarily dependent upon the physical magnitude (sound pressure) of the sound producing the sensation, but many other factors are influential. Sounds come in an infinite variety of frequencies, timbres, intensities, temporal patterns, and durations; each of these, as well as the characteristics of the individual listener and the context within which the sound is heard, has an influence on loudness.

Listening to a sound in the presence of another sound, which for the sake of simplicity we will call noise, results in the desired sound being, to some extent, less audible. This effect is called *masking*. If the noise is sufficiently loud, the signal can be completely masked, rendering it inaudible; at lower noise levels the signal will be partially masked, and only its apparent loudness may be reduced. If the desired sound is complex, it is possible for masking to affect only portions of the total sound. All this is dependent on the specific nature of both the signal and the masking sound.

Pitch is the subjective attribute of frequency, and while the basic correspondence between the two domains is obvious—low pitch to low frequencies and high pitch to high frequencies—the detailed relationships are anything but simple. Fortunately, waveforms that are periodic, however complex they may be, tend to be judged as having the same pitch as sine waves of the same repetition frequency. In other words, when a satisfactory pitch match has been made, the fundamental frequency of a complex periodic sound and a comparison sinusoid will normally be found to have the same frequency.

Sounds may be judged to have the same dimensions of loudness and pitch and yet sound very different from one another. This difference in sound quality, known as *timbre* in musical terminology, can relate to the tonal quality of sounds from specific musical instruments as they are played in live performance, to the character of tone imparted to all sounds processed through a system of recording and reproduction, and to the tonal modifications added by the architectural space within which the original performance or a reproduction takes place. Timbre is, therefore, a matter of fundamental importance in audio, since it can be affected by almost anything that occurs in the production, processing, storage, and reproduction of sounds. Timbre has many dimensions, not all of which have been fully identified or understood.

Chapter 2

Passive Circuit Components

Components used in electrical circuitry can be categorized into two broad classifications: *passive* or *active*. A voltage applied to a passive component results in the flow of current and the dissipation or storage of energy. Typical passive components include resistors, coils or inductors, and capacitors. For an example, the flow of current in a resistor results in radiation of heat; from a light bulb, the radiation of light as well as heat.

On the other hand, an active component either: 1) increases the level of electrical energy or 2) provides available electrical energy as a voltage. As an example of 1), an amplifier produces an increase in energy as a higher voltage or power level, while for 2), batteries and generators serve as energy sources.

Resistors

Resistors are components that have a nearly $0°$ phase shift between voltage and current over a wide range of frequencies, with the average value of resistance independent of the instantaneous value of voltage or current [1]. Preferred values of ratings are given American National Standards Institute (ANSI) standards or corresponding International Organization for Standardization (ISO) or military-grade (MIL) standards. Resistors are typically identified by their construction and by the resistance materials used. Fixed resistors have two or more terminals and are not adjustable. Variable resistors permit adjustment of resistance by a control handle or with a tool. A variety of common resistor types is shown in Figure 2.1.

Low-wattage fixed resistors are usually identified by color-coding on the body of the device, as illustrated in Figure 2.2. The four or five color bands represent a number or multiplier, with the last indicating the tolerance of the device. Sometimes a *precision* or *power* resistor may have the value stamped on it.

Resistors can be connected in series or in parallel with other resistors. For a series connection:

$$R_{total} = R_1 + R_2 + R_3 + R_n$$

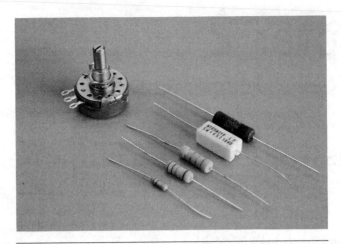

FIGURE 2.1 A selection of various common resistors

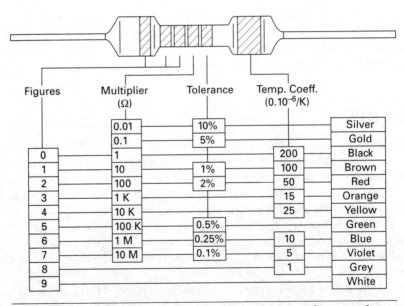

FIGURE 2.2 Color code for fixed resistors in accordance with International Electrotechnical Commission (IEC) publication 62. (*After* [2].)

where R_{total} is the resistance of the combination of resistors R_1, R_2, R_3, and R_n. The total resistance, thus, is greater than the largest single device in the string. In a series connection, the wattage rating of each resistor in the string should typically be the same.

For a parallel connection of two resistors:

$$R_{total} = \frac{R_1 \times R_2}{R_1 + R_2}$$

It follows that the total resistance of parallel resistors is smaller than the smallest of the devices. In the case where R1 and R2 are equal, the calculation can be simplified to:

$$R_{total} = R_1 \big/ 2$$

Wire-Wound Resistor

The resistance element of most wire-wound resistors is a resistance wire or ribbon wound as a single-layer helix over a ceramic or fiberglass core [1]. This form of construction causes wire-wound resistors to have a residual series inductance that affects phase shift at high frequencies, particularly in large-size devices. Wire-wound resistors have low noise and are stable with temperature, exhibiting temperature coefficients normally between ±5 and 200 ppm/°C. Resistance values between 0.1 and 100,000 Ω with accuracies between 0.001 and 20 percent are available, with power dissipation ratings between 1 and 250 W at 70°C. The resistance element is usually covered with a vitreous enamel, which can be molded in plastic. Special construction includes such items as enclosure in an aluminum casing for heatsink mounting or a special winding to reduce inductance. Resistor connections are made by self-leads or to terminals for other wires or printed circuit boards.

Metal Film Resistor

Metal film, or cermet, resistors have characteristics similar to wire-wound resistors except a much lower inductance [1]. They are commonly available as axial lead components in 1/8, 1/4, or 1/2 W ratings; in chip resistor form for high-density assemblies; or as resistor networks containing multiple resistors in one package suitable for printed circuit insertion, as well as in tubular form similar to high-power wire-wound resistors. Metal film resistors are essentially printed circuits using a thin layer of resistance alloy on a flat or tubular ceramic or other suitable insulating substrate. The shape and thickness of the conductor pattern determine the resistance value for each metal alloy used. Resistance is trimmed by cutting into part of the conductor pattern with an abrasive or a laser. Tin oxide is also used as a resistance material.

Carbon Film Resistor

Carbon film resistors are similar in construction and characteristics to axial lead metal film resistors [1]. Because the carbon film is a granular material, random noise may develop due to variations in the voltage drop between granules. This noise can be of sufficient level to affect the performance of circuits providing high gain when operating at low signal levels.

Carbon Composition Resistor

Carbon composition resistors contain a cylinder of carbon-based resistive material molded into a cylinder of high-temperature plastic, which also anchors the external leads [1]. These resistors can have noise problems similar to carbon film resistors, but their use in electronic equipment for the last 50 years has demonstrated their outstanding reliability, unmatched by most other components. These resistors are commonly available at values from 2.7 Ω with tolerances of 5, 10, and 20 percent in 1/8-, 1/4-, 1/2-, 1-, and 2-W sizes.

Control and Limiting Resistors

Resistors with a large negative temperature coefficient, *thermistors,* are often used to measure temperature, to limit inrush current into motors or power supplies, or to compensate bias circuits [1]. Resistors with a large positive temperature coefficient are used in circuits that have to match the coefficient of copper wire. Special-purpose resistors also include those that have a low resistance when cold and become a nearly open circuit when a critical temperature or current is exceeded to protect transformers or other devices.

Resistor Networks

Metal film or similar resistors are often packaged in a single module suitable for printed circuit mounting [1]. These devices see applications in digital circuits, as well as in fixed attenuators or padding networks.

Adjustable Resistors

Cylindrical wire-wound power resistors can be made adjustable with a metal clamp in contact with one or more turns not covered with enamel along an axial stripe [1]. Potentiometers are resistors with a movable arm that makes contact with a resistance element, which is connected to at least two other terminals at its ends. The resistance element can be circular or linear in shape, and often two or more sections are mechanically coupled or ganged for simultaneous control of two separate circuits. Resistance materials include all those described previously.

Trimmer potentiometers are similar in nature to conventional potentiometers except that adjustment requires a tool. Trimmers are available in single-turn and multiturn versions. The multiturn version offers greater adjustment accuracy.

Most potentiometers have a linear taper, which means that resistance changes linearly with control motion when measured between the movable arm and the "low," or counterclockwise, terminal. Audio gain controls, however, are often specified to have a logarithmic taper so that attenuation changes linearly in decibels. The resistance element of a potentiometer may also contain taps that permit the connection of other components as required in a specialized circuit.

Attenuators

Variable attenuators are adjustable resistor networks that show a calibrated increase in attenuation for each switched step [1]. For measurement of audio, video, and RF equipment, these steps may be decades of 0.1, 1, and 10 dB. Circuits for unbalanced and balanced fixed attenuators are shown in Figure 2.3. Fixed attenuator networks can

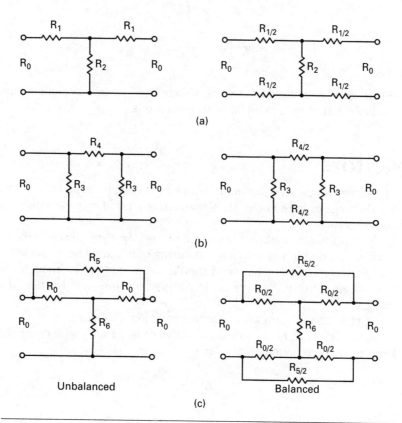

FIGURE 2.3   Unbalanced and balanced fixed attenuator networks for equal source and load resistance: (*a*) *T* configuration, (*b*) π configuration, (*c*) bridged-*T* configuration. (*After* [1].)

be cascaded and switched to provide step adjustment of attenuation inserted in a constant-impedance network.

Audio attenuators for professional applications generally are designed for a circuit impedance of 150 Ω or 600 Ω, although other impedances can be used for specific applications. Video attenuators are generally designed to operate with unbalanced 75-Ω grounded-shield coaxial cable. RF attenuators are designed for use with 75- or 50-Ω coaxial cable.

The basic properties of the attenuator networks shown in Figure 2.3 can be described as follows [3]:

$$A = \frac{\text{Input voltage}}{\text{Output voltage}} = \text{Attentuation}$$

$$\frac{R_1}{R_0} = \frac{R_0}{R_3} = \frac{A-1}{A+1}$$

$$\frac{R_2}{R_0} = \frac{R_0}{R_4} = \frac{2}{A-\left(\frac{1}{A}\right)}$$

$$\frac{R_5}{R_0} = \frac{R_0}{R_6} = A-1$$

Fortunately, attenuators of various values and for various applications are readily available without the need for any calculations.

Capacitors

Capacitors are passive components in which current leads voltage by nearly 90° over a wide range of frequencies [1]. Capacitors are rated by capacitance, voltage, materials, and construction.

A capacitor may have two voltage ratings. *Working voltage* is the normal operating voltage that should not be exceeded during operation. Use of the *test* or *forming voltage* stresses the capacitor and should occur only rarely in equipment operation. Good engineering practice is to use components only at a fraction of their maximum ratings.

A selection of common capacitors is shown in Figure 2.4.

Capacitors can be connected in series or in parallel with other capacitors. For a series connection of two devices:

$$C_{\text{total}} = \frac{C_1 \times C_2}{C_1 + C_2}$$

where C_{total} is the capacitance of the combination of capacitors C_1 and C_2. The total capacitance, thus, is less than the largest single device in the string.

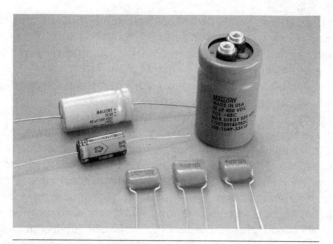

FIGURE 2.4   A variety of common capacitors

For a parallel connection of two capacitors:

$$C_{\text{total}} = C_1 + C_2$$

It follows that the total capacitance of parallel capacitors is greater than the largest of the devices.

Also of note are the following two characteristics (for initially discharged capacitors):

- The voltage applied to two capacitors connected in series is divided between the two capacitors inversely proportionally to their capacitance (thus, the larger part of the voltage is on the smaller capacitance).
- The current applied to two capacitors connected in parallel is divided between the two capacitors proportionally to their capacitance (thus, the larger part of the current goes through the larger capacitance).

Polarized Capacitors

Polarized capacitors can be used in only those applications where a positive sum of all DC and peak-AC voltages is applied to the positive capacitor terminal with respect to its negative terminal [1]. These capacitors include all tantalum and most aluminum electrolytic capacitors. These devices are commonly used in power supplies or other electronic equipment where these restrictions can be met.

Losses in capacitors occur because a practical capacitor has various resistances. These losses are usually measured as the *dissipation factor* at a frequency of 120 Hz.

Leakage resistance in parallel with the capacitor defines the time constant of discharge of a capacitor. This time constant can vary between a small fraction of a second and many hours, depending on capacitor construction, materials, and other electrical leakage paths, including surface contamination.

The *equivalent series resistance* of a capacitor is largely the resistance of the conductors of the capacitor plates and the resistance of the physical and chemical systems of the capacitor. When an alternating current is applied to the capacitor, the losses in the equivalent series resistance are the major causes of heat developed in the device. The same resistance also determines the maximum attenuation of a filter or bypass capacitor and the loss in a coupling capacitor connected to a load.

The *dielectric absorption* of a capacitor is the residual fraction of charge remaining in a capacitor after discharge. The residual voltage appearing at the capacitor terminals after discharge is of little concern in most applications, but can seriously affect the performance of converters in certain measurement applications.

The *self-inductance* of a capacitor determines the high-frequency impedance of the device and its ability to bypass high-frequency currents. The self-inductance is determined largely by capacitor construction and tends to be highest in common metal foil devices.

Nonpolarized Capacitors

Nonpolarized capacitors are used in circuits where there is no direct voltage bias across the capacitor [1]. They are also the capacitor of choice for most applications requiring capacity tolerances of 10 percent or less.

Film Capacitors

Plastic is a preferred dielectric material for capacitors because it can be manufactured with minimal imperfections in thin films [1]. A metal foil capacitor is constructed by winding layers of metal, plastic, metal, and plastic into a cylinder and then making a connection to the two layers of metal. A metallized foil capacitor uses two layers, each of which has a very thin layer of metal evaporated on one surface, thereby obtaining a higher capacity per volume in exchange for a higher equivalent series resistance. Metallized foil capacitors are self-repairing in the sense that the energy stored in the capacitor is often sufficient to burn away the metal layer surrounding a void in the plastic film.

Depending on the dielectric material and construction, capacitance tolerances between 1 and 20 percent are common, as are voltage ratings from 50 to 400 V. Construction types include axial lead capacitors with a plastic outer wrap, metal-encased units, and capacitors in a plastic enclosure suitable for printed wiring board insertion.

Polystyrene has the lowest dielectric absorption of 0.02 percent, a temperature coefficient of –20 to –100 ppm/°C, a temperature range to 85°C, and extremely low leakage. Capacitors between 0.001 and 2 µF can be obtained with tolerances from 0.1 to 10 percent.

Polycarbonate has an upper temperature limit of 100°C, with capacitance changes of about 2 percent up to this temperature. Polypropylene has an upper temperature limit of 85°C. These capacitors are particularly well suited for applications where high inrush currents occur, such as switching power supplies. Polyester is a low-cost material with an upper temperature limit of 125°C. Teflon and other high-temperature materials are used in critical applications.

Foil Capacitors

Mica capacitors are made of multiple layers of silvered mica packaged in epoxy or other plastic [1]. Available in tolerances of 1 to 20 percent in values from 10 to 10,000 pF, mica capacitors exhibit temperature coefficients as low as 100 ppm. Voltage ratings between 100 and 600 V are common. Mica capacitors are used mostly in high-frequency filter circuits where low loss and high stability are required.

Electrolytic Capacitors

Aluminum foil electrolytic capacitors can be made nonpolar through the use of two cathode foils in construction instead of anode and cathode foils [1]. With care in manufacturing, these capacitors can be produced with tolerance as tight as 10 percent at voltage ratings of 25 to 100 V peak. Typical values range from 1 to 1000 µF.

Ceramic Capacitors

Barium titanate and other ceramics have a high dielectric constant and a high breakdown voltage [1]. The exact formulation determines capacitor size, temperature range, and variation of capacitance over that range (and consequently capacitor application). Nonpolarized (NPO) rated capacitors range from 10 to 47,000 pF with a temperature coefficient of 0 to +30 ppm over a temperature range of –55 to +125°C.

Ceramic capacitors come in various shapes, the most common being the radial-lead disk. Multilayer monolithic construction results in small size, which exists both in radial-lead styles and as chip capacitors for direct surface mounting on a printed wiring board.

Polarized Capacitor Types and Construction

Polarized capacitors have a negative terminal—the cathode—and a positive terminal—the anode—and a liquid or gel between the two layers of conductors [1]. The actual dielectric is a thin oxide film on the cathode, which has been chemically roughened for maximum surface area. The oxide is formed with a *forming voltage,* higher than the normal operating voltage, applied to the capacitor during manufacture. The direct current flowing through the capacitor forms the oxide and also heats the capacitor.

Whenever an electrolytic capacitor is not used for a long period, some of the oxide film is degraded. It is reformed when voltage is applied again with a leakage current that decreases with time. Applying an excessive voltage to the capacitor causes a severe increase in leakage current, which can cause the electrolyte to boil. The resulting steam may escape by way of the rubber seal or may otherwise damage the capacitor. Application of a reverse voltage in excess of about 1.5 V will cause forming to begin on the unetched anode electrode. This can happen when pulse voltages superimposed on a DC voltage cause a momentary voltage reversal.

Aluminum Electrolytic Capacitors

Aluminum electrolytic capacitors use pure aluminum foil as electrodes, which are wound into a cylinder with an interlayer of paper or other porous material that contains the electrolyte [1]. (See Figure 2.5.) Aluminum ribbon staked to the foil at the minimum inductance location is brought through the insulator to the anode terminal, while the cathode foil is similarly connected to the aluminum case and cathode terminal.

Electrolytic capacitors typically have voltage ratings from 6.3 to 450 V and rated capacitances from 0.47 μF to several thousands of microfarads at the maximum voltage. Capacitance tolerance may range from ±20 to +80/−20 percent. The operating temperature range is often rated from −25 to +85°C or wider. Leakage current of an electrolytic capacitor may be rated as low as 0.002 times the capacity times the voltage rating to more than 10 times as much.

Tantalum Electrolytic Capacitors

Tantalum electrolytic capacitors are the capacitors of choice for applications requiring small size, 0.33- to 100-μF range at 10 to 20 percent tolerance, low equivalent series resistance, and low leakage current [1]. These devices are well suited where the less

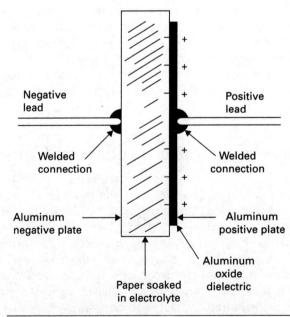

FIGURE 2.5 The basic design of an aluminum electrolytic capacitor. (*After* [4].)

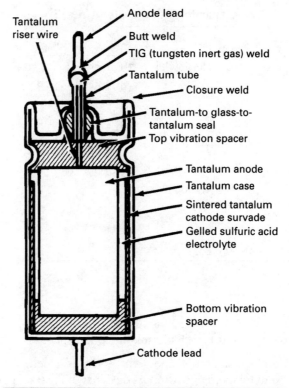

FIGURE 2.6   Basic construction of a tantalum capacitor. (*After* [4].)

costly aluminum electrolytic capacitors have performance issues. Tantalum capacitors are packaged in hermetically sealed metal tubes or with axial leads in epoxy plastic, as illustrated in Figure 2.6.

Identification

Capacitors of relatively large physical size usually have their values printed on the device with—at a minimum—the following information: 1) capacitance in microfarads (μF) or picofarads (pF), and 2) maximum working voltage. Some capacitors use a shorthand code to identify value, typically three digits followed by a letter, as follows:

- First digit of capacitor value
- Second digit of capacitor value
- Multiplier; multiply the first and second digits by the value given in Table 2.1
- Tolerance of the capacitor, as given in Table 2.1

TABLE 2.1 Capacitor Value Code (*After* [5])

Multiplier		Capacitor Tolerance		
Number	Multiply By	10 pF or less	Letter	More than 10 pF
0	1	± 0.1 pF	B	
1	10	± 0.25 pF	C	
2	100	± 0.5 pF	D	
3	1000	± 1.0 pF	F	± 1%
4	10,000	± 2.0 pF	G	± 2%
5	100,000		H	± 3%
			J	± 5%
8	0.01		K	± 10%
9	0.1		M	± 20%

For example:

"151K" = 15 × 10 = 150 pF, 10%
"759" = 75 × 0.1 = 7.5 pF (tolerance not specified)

Careful consideration should be given to selecting the capacitor working voltage for vacuum tube circuits. In most designs, the plate and heater voltages are applied simultaneously. For circuits that utilize silicon rectifiers, there is essentially no load on the power supply when it is first energized. As the tube heaters warm up, the load increases to its normal value. During this period, the voltage on any capacitor connected to the B+ line, whether directly or through a dropping resistor, will see something close to the maximum available output voltage of the supply. This being the case, such capacitors should be sized to accommodate the maximum voltage that the device will experience. Similar conditions can be observed with vacuum tube rectifiers, although the effect is less pronounced.

For circuits that are not subjected to high voltages under normal operating conditions, such as those in the grid or cathode circuit, consideration should be given to the possible consequences of an interelectrode short and the voltages that such an event could apply to devices in the circuit, notably capacitors. Because there is usually only a small cost delta between a 100 V capacitor of (for example) 0.1 µF and a device rated for 400 V, it usually makes sense to use the same (higher) rating for all devices in the circuit. The principal drawback to this approach is the increase in physical size of the device, which may present additional construction challenges.

Inductors and Transformers

Inductors are passive components in which voltage leads current by nearly 90° over a wide range of frequencies [1]. Inductors are usually coils of wire wound in the form of a cylinder. The current through each turn of wire creates a magnetic field that passes

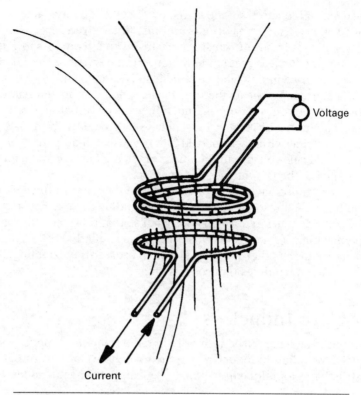

FIGURE 2.7   The basic principles of electromagnetic induction. (*After* [4].)

through every turn of wire in the coil. When the current changes, a voltage is induced in the wire and every other wire in the changing magnetic field. The voltage induced in the same wire that carries the changing current is determined by the inductance of the coil, and the voltage induced in the other wire is determined by the mutual inductance between the two coils. (See Figure 2.7.) A transformer has at least two coils of wire closely coupled by the common magnetic core, which contains most of the magnetic field within the transformer.

Inductors and transformers vary widely in size, weighing from less than 1 g to more than 1 ton, and have specifications ranging nearly as wide.

Losses in Inductors and Transformers

Inductors have resistive losses because of the resistance of the copper wire used to wind the coil [1]. An additional loss occurs because the changing magnetic field causes eddy currents to flow in every conductive material in the magnetic field. Using thin magnetic laminations or powdered magnetic material reduces these currents.

Losses in inductors are measured by the Q, or quality, factor of the coil at a test frequency. Losses in transformers are sometimes given as a specific insertion loss in decibels. Losses in power transformers are given as core loss in watts when there is no load connected and as regulation in percent, measured as the relative voltage drop for each secondary winding when a rated load is connected.

Transformer loss heats the transformer and raises its temperature. For this reason, power transformers are typically rated in watts or volt-amperes. The volt-ampere rating of a power transformer must always be greater than the DC power output from the rectifier circuit connected to it because volt-amperes, the product of the rms currents and rms voltages in the transformer, are larger by a factor of about 1.6 than the product of the DC voltages and currents.

Inductors also have capacitance between the wires of the coil, which causes the coil to have a self-resonance between the winding capacitance and the self-inductance of the coil. Circuits are normally designed so that this resonance is outside of the frequency range of interest. Transformers are similarly limited. They also have capacitance to the other winding(s), which causes stray coupling. An electrostatic shield between windings reduces this effect.

Air-Core Inductors

Air-core inductors are used primarily in radio frequency applications because of the need for values of inductance in the microhenry or lower range [1]. The usual construction is a multilayer coil made self-supporting with adhesive-covered wire.

Ferromagnetic Cores

Ferromagnetic materials have a permeability much higher than air or vacuum and cause a proportionally higher inductance of a coil that has all its magnetic flux in this material [1]. Ferromagnetic materials in audio and power transformers or inductors usually are made of silicon steel laminations stamped in the forms of letters E or I, as illustrated in Figure 2.8. At higher frequencies, powdered ferric oxide is used. The continued magnetization and remagnetization of silicon steel and similar materials in opposite directions does not follow the same path in both directions, but encloses an area in the magnetization curve and causes a hysteresis loss at each pass, or twice per AC cycle.

Another simple classic geometry is the toroid [1]. This doughnut-shaped winding (Figure 2.9) is a close relative of the long solenoid. Indeed, it may be viewed as a long solenoid whose ends have been joined.

Most devices used in vacuum tube applications utilize multiple secondary windings. At minimum a filament winding providing 6.3 V AC and a plate supply winding providing 200 to 400 V AC are typically provided.

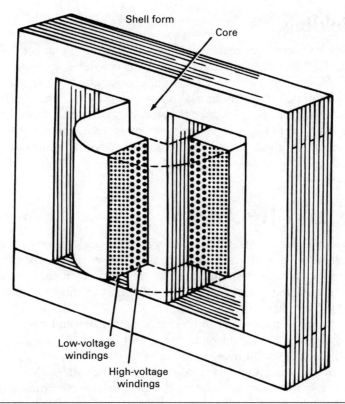

FIGURE 2.8   Physical construction of an E-shaped core transformer. The low- and high-voltage windings are stacked as shown. (*After* [4].)

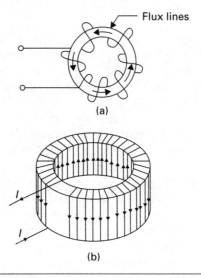

FIGURE 2.9   Physical construction of a toroid transformer: (*a*) a toroid inductor, (*b*) a closely wound toroidal coil. (*After* [6].)

Shielding

Transformers and coils radiate magnetic fields that can induce voltages in other nearby circuits [1]. Similarly, coils and transformers can develop voltages in their windings when subjected to magnetic fields from another transformer, motor, or power circuit. Steel mounting frames or chassis conduct these fields, offering less reluctance than air.

The simplest way to reduce the stray magnetic field from a power transformer is to wrap a copper strip as wide as the coil of wire around the transformer enclosing all three legs of the core. Shielding occurs by having a short circuit turn in the stray magnetic field outside of the core.

Diodes and Rectifiers

A diode is a passive electronic device that has a positive anode terminal and a negative cathode terminal and a nonlinear voltage-current characteristic [1]. A rectifier is assembled from one or more diodes for the purpose of obtaining a direct current from an alternating current; this term also refers to large diodes used for this purpose. Many types of diodes exist.

Over the years, a great number of constructions and materials have been used as diodes and rectifiers. Semiconductor materials such as germanium, silicon, selenium, copper oxide, or gallium arsenide can be processed to form a positive-negative (*pn*) junction that has a nonlinear diode characteristic. Although all these systems of rectification have seen use, the most widely used rectifier in electronic equipment is the silicon diode. The remainder of this section deals only with these and other silicon two-terminal devices.

The pn Junction

When biased in a reverse direction at a voltage well below breakdown, the diode reverse current is composed of two currents [1]. One current is caused by leakage due to contamination and is proportional to voltage. The intrinsic diode reverse current is independent of voltage but doubles for every 10°C in temperature (approximately). The forward current of a silicon diode is approximately equal to the leakage current multiplied by e (= 2.718) raised to the power given by the ratio of forward voltage divided by 26 mV with the junction at room temperature. In practical rectifier calculations, the reverse current is considered to be important in only those cases where a capacitor must hold a charge for a time, and the forward voltage drop is assumed to be constant at 0.7 V, unless a wide range of currents must be considered.

All diode junctions have a junction capacitance that is approximately inversely proportional to the square of the applied reverse voltage. This capacitance rises further with applied forward voltage. When a rectifier carries current in a forward direction, the junction capacitance builds up a charge. When the voltage reverses across the junction, this charge must flow out of the junction, which now has a lower capacitance, giving rise to a current spike in the opposite direction of the forward current. After the reverse-recovery time, this spike ends, but interference may be radiated into

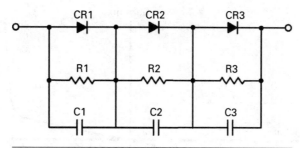

FIGURE 2.10 A high-voltage rectifier stack.
(*After* [4].)

low-level circuits. For this reason, rectifier diodes are sometimes bypassed with capacitors of about 0.1 µF located close to the diodes. Rectifiers used in high-voltage assemblies use bypass capacitors and high-value resistors to reduce noise and equalize the voltage distribution across the individual diodes (see Figure 2.10).

Tuning diodes have a controlled reverse capacitance that varies with applied direct tuning voltage. This capacitance may vary over a 2-to-1 to as high as a 10-to-1 range and is used to change the resonant frequency of tuned RF circuits. These diodes find application in radio and television receiver circuits.

Zener Diodes and Reverse Breakdown

When the reverse voltage on a diode is increased to a certain critical voltage, the reverse leakage current will increase rapidly, or *avalanche* [1]. This breakdown, or *zener* voltage, sets the upper voltage limit a rectifier can experience in normal operation because the peak reverse currents may become as high as the forward currents. Rectifier and other diodes have a rated peak reverse voltage, and some rectifier circuits may depend on this reverse breakdown to limit high-voltage spikes that enter equipment from the power line. It should also be noted that diode dissipation is very high during these periods.

The reverse breakdown voltage can be controlled in manufacture to a few percent and used to advantage in a class of devices known as zener diodes, used extensively in voltage-regulator circuits. Voltage-regulator diodes are available in voltages from 2.4 to 200 V with rated dissipation typically between 1/4 and 10 W. The forward characteristics of a zener diode usually are not specified, but are similar to those of a conventional diode.

Varistor

Varistors are symmetrical nonlinear voltage-dependent resistors, behaving not unlike two zener diodes connected back to back [1]. The current in a varistor is proportional to applied voltage raised to a power N. These devices are normally made of zinc oxide, which can be produced to have an N factor of 12 to 40. In circuits at normal operating voltages, varistors are nearly open circuits shunted by a capacitor of a few hundred

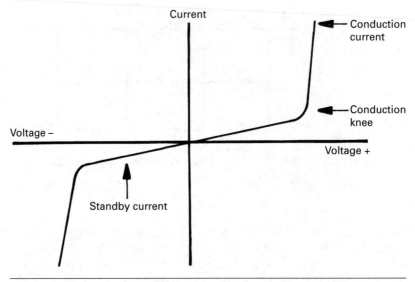

FIGURE 2.11 The current versus voltage transfer curve for a varistor. Note the *conduction knee*. (*After* [4].)

to a few thousand picofarads. Upon application of a high-voltage pulse, such as a lightning discharge, they conduct a large current, thereby "absorbing" the pulse energy in the bulk of the material[1] with only a relatively small increase in voltage, thus protecting the circuit. (See Figure 2.11.) Varistors are available for operating voltages from 10 to 1000 V rms and can handle pulse energies from 0.1 to more than 100 J and maximum peak currents from 20 to 2000 A. Typical applications include protection of power supplies and power-switching circuits, and the protection of telephone and data-communication lines.

Indicators

Indicators are generally passive components that send a message to the operator of the equipment [1]. This message is most commonly a silent visual indication that the equipment is operating in some particular mode, is ready to operate, or is not ready. Indicator lights of different colors illuminating a legend or having an adjacent legend are most commonly used.

Miniature light bulbs are incandescent devices operating at low voltage between 6 and 24 V. The resistance of the filament increases with temperature, varying by as much as a factor of 16 from cold to hot.

Solid-state lamps or light-emitting diodes (LED) are pn-junction lasers that generate light when diode current exceeds a critical threshold value. Visible red light is emitted

---

[1] Dissipated as heat.

from gallium arsenide phosphide junctions. Green or amber light is emitted from doped gallium phosphide junctions. The junctions have a forward voltage drop of 1.7 to 2.2 V at a normal operating current of 10 to 50 mA. Other visible colors are commercially available. The LED is encased singly in round or rectangular plastic cases or assembled as multiples.

Electrons emitted from a heated cathode or a cold cathode can cause molecules of low-pressure gas, such as neon, to ionize and to emit light. Neon lamps require a current-limited supply of at least 90 V to emit orange light. The most frequent use of neon lamps is to indicate the presence of power line voltage. By means of a series resistor, the current is limited to a permissible value.

Electrical Conductors

At the heart of any electrical system is the cable used to tie distant parts together [7]. Conductors are rated by the American Wire Gauge (AWG) scale. The smallest is no. 36, and the largest is no. 000. There are 40 sizes in between. Sizes larger than no. 0000 AWG are specified in thousand circular mil units, referred to as "MCM" units (M is the Roman numeral expression for 1,000). The cross-sectional area of a conductor doubles with each increase of three AWG sizes. The diameter doubles with every six AWG sizes.

Most conductors used for signal and power distribution are made of copper. Stranded conductors are used where flexibility is required. Stranded cables usually are more durable than solid conductor cables of the same AWG size.

Resistance and inductance are the basic electrical parameters of concern in the selection of wire for electronic systems. Resistivity is commonly measured in ohm-centimeters (Ω-cm). Table 2.2 lists the resistivity of several common materials.

Ampacity is the measure of the ability of a conductor to carry electric current. Although all metals will conduct current to some extent, certain metals are more efficient than others. The three most common high-conductivity metals are

- Silver, with a resistivity of 9.8 Ω/circular mil-foot
- Copper, with a resistivity of 10.4 Ω/cmil-ft
- Aluminum, with a resistivity of 17.0 Ω/cmil-ft

TABLE 2.2   Resistivity of Common Materials

Material	Resistivity ($\mu\,\Omega$-cm)
Silver	1.468
Copper	1.724
Aluminum	2.828
Steel	5.88
Brass	7.5
Iron	9.8

The ampacity of a conductor is determined by the type of material used, the cross-sectional area, and the heat-dissipation effects of the operating environment.

Effects of Inductance

Current through a wire results in a magnetic field [7]. All magnetic fields store energy, and this energy cannot be changed in zero time. Any change in the field takes a finite length of time to occur. Inductance (H) is the property of opposition to changes in energy level. The inductance of equipment interconnection cable is usually a distributed parameter.

Voltage drop in a conductor is a function of resistance and inductance. The *skin effect* and circuit geometry affect both parameters.

Skin Effect

The effective resistance offered by a given conductor to radio frequencies is considerably higher than the ohmic resistance measured with direct current [8]. This is because of an action known as the skin effect, which causes the currents to be concentrated in certain parts of the conductor and leaves the remainder of the cross-section to contribute little or nothing toward carrying the applied current.

When a conductor carries an alternating current, a magnetic field is produced that surrounds the wire. This field continually expands and contracts as the AC wave increases from zero to its maximum positive value and back to zero, then through its negative half-cycle. The changing magnetic lines of force cutting the conductor induce a voltage in the conductor in a direction that tends to retard the normal flow of current in the wire. This effect is more pronounced at the center of the conductor. Thus, current within the conductor tends to flow more easily toward the surface of the wire. The higher the frequency, the greater the tendency for current to flow at the surface. The depth of current flow is a function of frequency, and it is determined from:

$$d = \frac{2.6}{\sqrt{\mu \times f}}$$

where d = depth of current in mils, μ = permeability (copper = 1, steel = 300), and f = frequency of signal in MHz.

It can be calculated that at a frequency of 100 kHz, current flow penetrates a conductor by 8 mils. At 1 MHz, the skin effect causes current to travel in only the top 2.6 mils in copper, and even less in almost all other conductors. Therefore, the series impedance of conductors at high frequencies is significantly higher than at low frequencies.

Coaxial Cable

The motion of electrical energy requires the presence of an electric field and a magnetic field [7]. Any two conductors can direct the flow of energy. The basic geometry for energy transport is two parallel conductors, as illustrated in Figure 2.12.

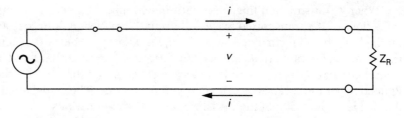

FIGURE 2.12   Basic transmission line circuit. (*After* [7].)

The transmission line exhibits distributed capacitance C and distributed inductance L along its length. When the switch in the diagram is closed, current begins to flow, charging the capacitance. This current also establishes a magnetic field around both conductors. The energy in these two fields is supplied at a fixed rate. The voltage wave propagates down the line at a fixed velocity, given by:

$$V = \sqrt{L \times C}$$

The velocity in the conductors is typically about one-half the speed of light.

Energy is stored on the line, and as energy is added, it must be transported past any existing storage. This requires an electric field and a magnetic field behind the wavefront. The current I that flows in the line is given by:

$$I = \frac{V}{\sqrt{\frac{L}{C}}}$$

$$I = \frac{V}{Z}$$

where Z = the characteristic impedance of the line in ohms.

If the transmission line were cut at some point and terminated in an impedance Z, energy would continue to flow on the line as if it had infinite length. When the wavefront reaches the termination, energy is dissipated per unit time rather than being stored per unit time.

The transmission line principles presented here represent an ideal circuit. In a practical transmission line, many factors contribute to losses and some radiation, including the following:

- Skin effect
- Dielectric and conductive losses
- Irregularities in geometry

These factors change with the frequency of the transported wave. For audio applications with relatively short-length runs, these issues can be largely ignored.

When a transmission line is not terminated in its characteristic impedance, reflections of the transported wave will occur. When a signal reaches an open circuit on the line, the total current flow at the open point must be zero. A reflected wave, therefore, is generated that cancels this current. If, on the other hand, a signal reaches a short circuit on the line, a reflected wave is generated that cancels the voltage. Reflections of these types return energy to the source. The signal at any point along the line is a composite of the initial signal and any reflections.

Sinusoidal signals are assumed when the input impedance of a transmission line is discussed. The input impedance is determined by the following:

- Characteristic impedance of the line
- Terminating impedance
- Applied frequency
- Length of the line

Reflected energy reaching the source modifies the voltage-current relationship. On short unterminated lines, the input impedance can vary significantly. If the reflected wave returns in phase with the input signal, no current will flow; the input impedance is infinite. If the input signal returns 90° out of phase, the line will appear as a pure reactive load to the source.

Operating Principles

A coaxial transmission line consists of concentric center and outer conductors that are separated by a dielectric material [7]. When current flows along the center conductor, it establishes an electric field. The electric flux density and the electric field intensity are determined by the dielectric constant of the dielectric material. The dielectric material becomes polarized with positive charges on one side and negative charges on the opposite side. The dielectric, therefore, acts as a capacitor with a given capacitance per unit length of line. Properties of the field also establish a given inductance per unit length and a given series resistance per unit length. If the transmission line resistance is negligible and the line is terminated properly, the following formula describes the characteristic impedance Z_0 of the cable:

$$Z_0 = \sqrt{\frac{L}{C}}$$

where L = inductance in H/ft and C = capacitance in F/ft.

Coaxial cables typically are manufactured with 50Ω or 75Ω characteristic impedances. Other characteristic impedances are possible by changing the diameter of the center and outer conductors. Figure 2.13 illustrates the relationship between characteristic impedance and the physical dimensions of the cable.

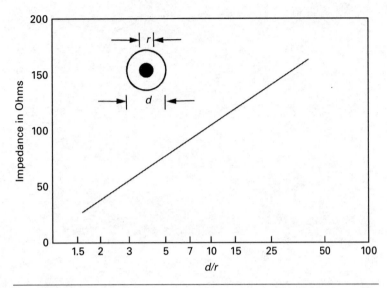

FIGURE 2.13 The interdependence of coaxial cable physical dimensions and characteristic impedance. (*After* [7].)

Cables for Audio Applications

A variation of coaxial cable as described in the previous section is typically used for low-level audio applications. A center conductor with a wrapped outer shield carries the common (ground) connection and serves to minimize induced signals (noise) in the signal-carrying conductor. A wide variety of such cables is available, with most distinctions based on the type of outer conductor (shielding) used.

For long runs in electrically noisy environments, single-ended (signal + common) connections, even with shielded cable, may not be the best approach. An alternative used widely in professional applications is a balanced circuit where the signal being transported is not referenced to ground. This approach is discussed in some detail in Chapter 5. The cabling can be as simple as an unshielded twisted pair. For most consumer audio applications where the interconnecting cables are relatively short (12 ft or less), single-ended interconnections are typically used with good results.

Apart from these cable types, wire used in construction of receivers and audio amplifiers can be loosely categorized as "hookup wire." The choices here are wide and varied, and include the following:

- Wire size
- Jacket material
- Solid or stranded conductor
- Color

For any sizeable system, some form of color coding is advisable. For example, high-voltage circuits might all be connected using red-colored wire, grounds black wire, signal lines yellow, and so on.

Chapter 3

Vacuum Tube Principles

Vacuum tubes are the active components that enabled the amplification and control of audio, radio frequency, and other signals and helped bring about the growth of the electronics industry from a laboratory curiosity early in the twentieth century to a high state of maturity in the 1960s and beyond.

The vacuum tube is an ingenious device. A heated cathode coated with rare-earth oxides in a vacuum causes a cloud of electrons to exist near the cathode. A positive anode voltage with respect to the cathode causes some of these electrons to flow as a current to the anode. A grid of wires at a location between anode and cathode and biased at a control voltage with respect to the cathode causes a greater or lesser amount of anode current to flow. Other intervening grids also control the anode current and, if biased with a positive voltage, draw grid current from the total cathode current. The vacuum tube uses this flow of free electrons to produce useful work. The physical shape and location of the grids relative to the plate and cathode are the main factors that determine the *amplification factor* (μ) and other parameters of the device. The physical size and types of material used to construct the individual elements determine the power capability of the tube. A wide variety of tube designs are available to commercial and industrial users. By far the most common are triodes, tetrodes, and pentodes.

The importance of the electron tube lies in its ability to control almost instantly the flight of the millions of electrons supplied by the cathode [1]. It accomplishes this control with a minimum of energy. Because it is almost instantaneous in this action, the electron tube can operate efficiently and accurately at high frequencies.

Characteristics of Electrons

Electrons are minute, negatively charged particles that are constituents of all matter [2]. They have a mass of 9×10^{-28} g (1/1840 that of a hydrogen atom) and a charge of 1.59×10^{-19} coulomb. Electrons are always identical, irrespective of their source. Atoms are composed of one or more such electrons associated with a much heavier nucleus, which has a positive charge equal to the number of the negatively charged electrons contained in the atom; an atom with a full quota of electrons is electrically neutral. The differences in chemical elements arise from differences in the nucleus and in the number of associated electrons.

Free electrons can be produced in a number of ways [3]. *Thermonic emission* is the method normally employed in vacuum tubes. The principle of thermonic emission states that if a solid body is heated sufficiently, some of the electrons that it contains will escape from the surface into the surrounding space. Electrons also are ejected from solid materials as a result of the impact of rapidly moving electrons or ions. This phenomenon is referred to as *secondary electron emission*, because it is necessary to have a primary source of electrons (or ions) before the secondary emission can be obtained. Finally, it is possible to pull electrons directly out of solid substances by an intense electrostatic field at the surface of the material.

Positive ions represent atoms or molecules that have lost one or more electrons and so have become charged bodies having the weight of the atom or molecule concerned and a positive charge equal to the negative charge of the lost electrons. Unlike electrons, positive ions are not all alike and may differ in charge, weight, or both. They are much heavier than electrons and resemble the molecule or atom from which they are derived. Ions are designated according to their origin, such as mercury ions or hydrogen ions.

Electron Optics

Electrons and ions are charged particles and, as such, have forces exerted upon them by an electrostatic field in the same way as other charged bodies [3]. Electrons, being negatively charged, tend to travel toward the positive or anode electrode, while the positively charged ions travel in the opposite direction (toward the negative or cathode electrode). The force F exerted upon a charged particle by an electrostatic field is proportional to the product of the charge e of the particle and the voltage gradient G of the electrostatic field [3]:

$$F = G \times e \times 10^7$$

where F = force in dynes, G = voltage gradient in volts per centimeter, and e = charge in coulombs.

This force upon the ion or electron is exerted in the direction of the electrostatic flux lines at the point where the charge is located. The force acts toward or away from the positive terminal, depending upon whether a negative or positive charge, respectively, is involved.

The force that the field exerts on the charged particle causes an acceleration in the direction of the field at a rate that can be calculated by the laws of mechanics where the velocity does not approach that of light:

$$A = \frac{F}{m}$$

where A = acceleration in centimeters per second, F = force in dynes, and m = mass in grams.

The velocity an electron or ion acquires in being acted upon by an electrostatic field can be expressed in terms of the voltage through which the electron (or ion) has fallen in acquiring the velocity. For velocities well below the speed of light, the relationship between velocity and the acceleration voltage is:

$$v = \sqrt{\frac{2 \times V \times e \times 10^7}{m}}$$

where:

v = velocity in centimeters per second corresponding to V

V = accelerating voltage

e = charge in coulombs

m = mass in grams

Electrons and ions move at great velocities in even moderate-strength fields. For example, an electron dropping through a potential difference of 2500 V will achieve a velocity of approximately one-tenth the speed of light.

Magnetic Field Effects

An electron in motion represents an electric current of magnitude ev, where e is the magnitude of the charge on the electron and v is its velocity [3]. A magnetic field accordingly exerts a force on a moving electron in a manner similar to the force it exerts on an electric current in a wire. The magnitude of the force is proportional to the product of the equivalent current ev represented by the moving electron and the strength of the component of the magnetic field in a direction at right angles to the motion of the electron. The resulting force is, then, in a direction at right angles both to the direction of motion of the electron and to the component of the magnetic field that is producing the force. As a result, an electron entering a magnetic field with a high velocity will follow a curved path. Because the acceleration of the electron that the force of the magnetic field produces is always at right angles to the direction in which the electron is traveling, an electron moving in a uniform magnetic field will follow a circular path. The radius of this circle is determined by the strength of the magnetic field and the speed of the electron moving through the field.

When an electron is subjected to the simultaneous action of both electric and magnetic fields, the resulting force acting on the electron is the vector sum of the force resulting from the electric field and the force resulting from the magnetic field, each considered separately.

Magnetic fields are not used for receiving tubes. Microwave power tubes, on the other hand, use magnetic fields to confine and focus the electron stream. Cathode ray tubes utilize magnetic fields for deflecting a stream of electrons across the face of the device.

Thermal Emission from Metals

Thermonic emission is the phenomenon of an electric current leaving the surface of a material as the result of thermal activation [3]. Electrons with sufficient thermal energy to overcome the surface-potential barrier escape from the surface of the material. This thermally emitted electron current increases with temperature because more electrons have sufficient energy to leave the material.

The number of electrons released per unit area of an emitting surface is related to the absolute temperature of the emitting material and a quantity b that is a measure of the work an electron must perform in escaping through the surface, according to the following equation [3]:

$$I = AT^2 \varepsilon^{-b/T}$$

where:

T = absolute temperature of the emitting material

b = the work an electron must perform in escaping the emitter surface

I = electron current in amperes per square centimeter

A = a constant (value varies with type of emitter)

The exponential term in the equation accounts for most of the variation in emission with temperature. The temperature at which the electron current becomes appreciable is accordingly determined almost solely by the quantity b. Figure 3.1 plots the emission resulting from a cathode operated at various temperatures.

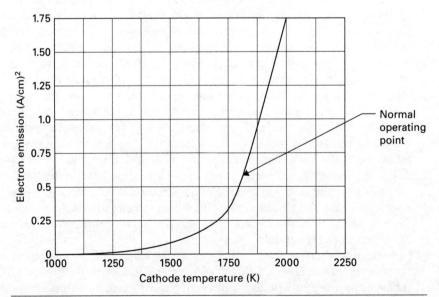

FIGURE 3.1   Variation of electron emission as a function of absolute temperature for a thoriated-tungsten emitter. (*After* [3].)

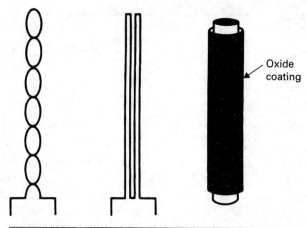

FIGURE 3.2   Common types of heater and cathode structures. (*After* [4].)

Thermal electron emission can be increased by applying an electric field to the cathode. This field lowers the surface-potential barrier, enabling more electrons to escape. This field-assisted emission is known as the *Schottky effect*.

Figure 3.2 illustrates common heater-cathode structures.

Secondary Emission

Almost all metals, and some insulators, will emit low-energy electrons (secondary electrons) when bombarded by other energetic electrons [3]. The number of secondary electrons emitted per primary electron is determined by the velocity of the primary bombarding electrons and the nature and condition of the material composing the surface being bombarded. Figure 3.3 illustrates a typical relationship for two types of surfaces. As shown in the figure, no secondary electrons are produced when the primary velocity is low. However, with increasing potential (and consequently higher velocity), the ratio of secondary to primary electrons increases, reaching a maximum and then decreasing. With pure-metal surfaces, the maximum ratio of secondary to primary electrons ranges from less than 1 to approximately 3.

The majority of secondary electrons emitted from a conductive surface have relatively low velocity. However, a few secondary electrons usually are emitted with a velocity nearly equal to the velocity of the bombarding primary electrons.

For insulators, the ratio of secondary to primary electrons as a function of primary electron potential follows along the same lines as for metals. The net potential of the insulating surface being bombarded is affected by the bombardment. If the ratio of secondary to primary current is less than unity, the insulator acquires a net negative charge because more electrons arrive than depart. This causes the insulator to become more negative and, finally, to repel most of the primary electrons, resulting in a blocking action. In the opposite extreme, when the ratio of secondary to primary

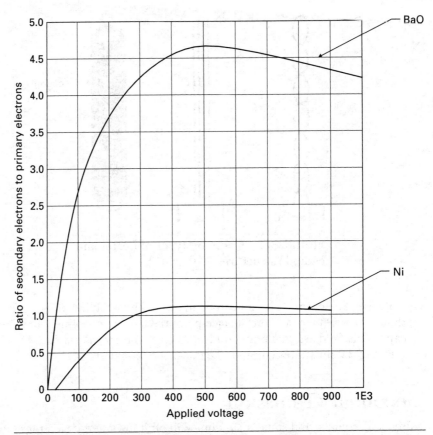

FIGURE 3.3 Ratio of secondary emission current to primary current as a function of primary electron velocity. (*After* [3].)

electrons exceeds unity, the insulating surface loses electrons through secondary emission faster than they arrive; the surface becomes increasingly positive. This action continues until the surface is sufficiently positive that the ratio of secondary to primary electrons decreases to unity as a result of the increase in the velocity of the bombarding electrons, or until the surface is sufficiently positive that it attracts back into itself a significant number of secondary electrons. This process makes the number of electrons gained from all sources equal to the number of secondary electrons emitted.

Types of Vacuum Tubes

Electrons are of no value in a vacuum tube unless they can be put to work [1]. A tube is therefore designed with the elements necessary to utilize electrons as well as those required to produce them. These elements consist of a cathode and one or more supplementary electrodes. The electrodes are enclosed in an evacuated envelope having the necessary connections brought out through airtight seals. The air is removed from the envelope to allow free movement of the electrons and to prevent damage to the emitting surface of the cathode.

Diode

A diode is a two-electrode vacuum tube containing a cathode, which emits electrons by thermonic emission, surrounded by an anode (or plate) [3] (see Figure 3.4). Such a tube is inherently a rectifier, because when the anode is positive, it attracts electrons; current, therefore, passes through the tube. When the anode is negative, it repels the electrons and no current flows.

The typical relationship between anode voltage and current flowing to the positive anode is shown in Figure 3.5. When the anode voltage is sufficiently high, electrons are drawn from the cathode as rapidly as they are emitted. The anode current is then limited by the electron emission of the cathode and, therefore, depends upon cathode temperature rather than anode voltage.

At low anode voltages, however, plate current is less than the emission of which the cathode is capable. This occurs because the number of electrons in transit between the cathode and plate at any instant cannot exceed the number that will produce a negative *space charge,* which completely neutralizes the attraction of the positive plate upon the electrons just leaving the cathode. All electrons in excess of the number necessary to neutralize the effects of the plate voltage are repelled into the cathode by the negative space charge of the electrons in transit; this situation applies irrespective of how many excess electrons the cathode emits. When the plate current is limited in this way by space charge, plate current is determined by plate potential and is substantially independent of the electron emission of the cathode.

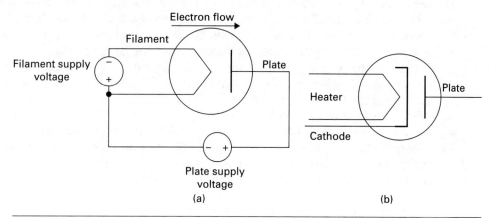

FIGURE 3.4   Vacuum diode: (*a*) directly heated cathode, (*b*) indirectly heated cathode. (*After* [4].)

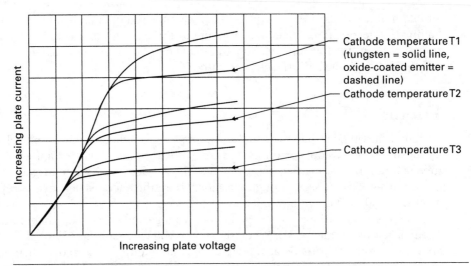

FIGURE 3.5 Anode current as a function of anode voltage in a two-electrode tube for three cathode temperatures. (*After* [3].)

Detailed examination of the space-charge situation will reveal that the negative charge of the electrons in transit between the cathode and the plate is sufficient to give the space in the immediate vicinity of the cathode a slight negative potential with respect to the cathode. The electrons emitted from the cathode are projected out into this field with varying emission velocities. The negative field next to the cathode causes the emitted electrons to slow as they move away from the cathode, and those having a low velocity of emission are driven back into the cathode. Only those electrons having the highest velocities of emission will penetrate the negative field near the cathode and reach the region where they are drawn toward the positive plate. The remainder (those electrons having low emission velocities) will be brought to a stop by the negative field adjacent to the cathode and will fall back into the cathode.

The energy that is delivered to the tube by the source of anode voltage is first expended in accelerating the electrons traveling from the cathode to the anode; it is converted into kinetic energy. When these swiftly moving electrons strike the anode, this kinetic energy is then transformed into heat as a result of the impact and appears at the anode in the form of heat that must be radiated to the walls of the tube.

Diodes are commonly used as rectifiers in AC-powered systems to convert the AC supply voltage to the DC voltage(s) necessary for other devices in the unit [1]. Figure 3.6*a* shows the basic diode circuit, and Figure 3.6*b* illustrates the rectified output current produced by an alternating input voltage.

Rectifier tubes are commonly available in two primary configurations:

- One plate and one cathode for half-wave rectifier applications
- Two plates and one or two cathodes in the same device for full-wave rectifier applications

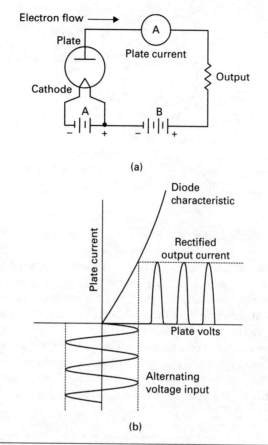

FIGURE 3.6 Basic diode circuit: (*a*) operating circuit, (*b*) current characteristics of a rectifier circuit. (*After* [1].)

Example Device

The 5U4 is a full-wave rectifier used in the power supplies of tube-based amplifiers and receivers [1]. The device uses an eight-pin (octal) mounting base. The coated filament is designed to operate from the AC line through a step-down transformer. The pin arrangement of the 5U4 is shown in Figure 3.7. The rating chart is shown in Figure 3.8. Selected operating values are listed in Table 3.1.

Triode

The triode is a three-element device commonly used in a wide variety of systems [5]. Triodes have three internal elements: the cathode, control grid, and plate. Most tubes are cylindrically symmetrical. The filament or cathode structure, the grid, and the

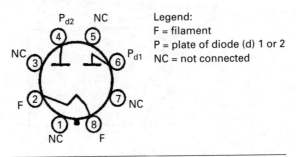

FIGURE 3.7 Pin arrangement of the 5U4GB. (*After* [1].)

anode are usually cylindrical in shape and are mounted with the axis of each cylinder along the center line of the tube, as illustrated in Figure 3.9.

The grid normally is operated at a negative potential with respect to the cathode, and so attracts no electrons [3]. However, the extent to which it is negative affects the electrostatic field in the vicinity of the cathode and, therefore, controls the number of electrons that pass between the grid and the plate. The grid, in effect, functions as an imperfect electrostatic shield. It allows some, but not all, of the electrostatic flux from the anode to leak between its wires. The number of electrons that reach the anode in a triode tube under space charge–limited conditions is determined almost solely by the electrostatic field near the cathode; the field in the rest of the interelectrode space has little effect. This phenomenon results because the electrons near the cathode

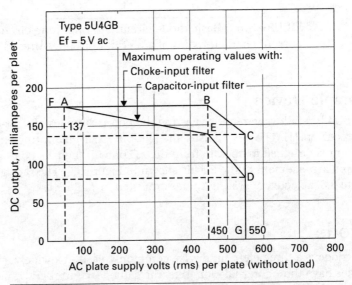

FIGURE 3.8 Rating chart of the 5U4GB. (*After* [1].)

TABLE 3.1   Key Parameters for the 5U4GB as a Full-Wave Rectifier (*After* [1].)

Maximum Ratings (Design Maximum Values)

Peak inverse plate voltage	1550 V
Peak plate current (per plate)	1 A
AC plate supply voltage (per plate, rms)	See Figure 3.8
Average output current (per plate)	See Figure 3.8
Heater (AC)	5 V, 3A

Values for Operation with Capacitor Input to Filter

AC plate-to-plate supply voltage (rms)		600 V	900 V	1100 V
Filter input capacitor value[1]		40 µF	40 µF	40 µF
Total effective plate supply impedance per plate		21 Ω	67 Ω	97 Ω
DC output voltage at input to filter (approx.) at half-load current of:	150 mA	335 V	—	—
	137.5 mA	—	520 V	—
	81 mA	—	—	680 V
DC output voltage at input to filter (approx.) at full-load current of:	300 mA	290 V	—	—
	275 mA	—	460 V	—
	162 mA	—	—	630 V
Voltage regulation (approx.), half-load to full-load current		45 V	60 V	50 V

Values for Operation with Choke Input to Filter

AC plate-to-plate supply voltage (rms)		900 V	1100 V
Filter input choke value		10 H	10 H
DC output voltage at input to filter (approx.) at half-load current of:	174 mA	335 V	—
	137.5 mA	—	455 V
DC output voltage at input to filter (approx.) at full-load current of:	348 mA	340 V	—
	275 mA	—	440 V
Voltage regulation (approx.), half-load to full-load current		15 V	15 V

[1] Higher values of capacitance than indicated may be used, but the effective plate-supply impedance may have to be increased to prevent exceeding the maximum rating for peak plate current.

are moving slowly compared with the electrons that have traveled some distance toward the plate. The result of this condition is that the volume density of electrons in proportion to the rate of flow is large near the cathode and low in the remainder of the interelectrode space. The total space charge of the electrons in transit toward the plate, therefore, consists almost solely of the electrons in the immediate vicinity

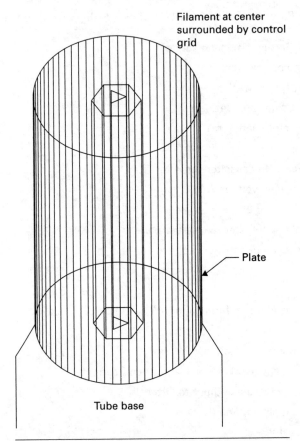

Filament at center
surrounded by control
grid

Plate

Tube base

FIGURE 3.9 Mechanical configuration of a
triode. (*After* [5].)

of the cathode. After an electron has traveled beyond this region, it reaches the
plate so quickly as to contribute to the space charge for only a brief additional time
interval. The result is that the space current in a three-electrode vacuum tube is, for
all practical purposes, determined by the electrostatic field that the combined action
of the grid and plate potentials produces near the cathode.

When the grid structure is symmetrical, the field E at the surface of the cathode is
proportional to the quantity

$$E_c + \frac{E_b}{\mu}$$

where E_c = control grid voltage (with respect to cathode), E_b = anode voltage (with
respect to cathode), and μ = a constant determined by the geometry of the tube.

The constant μ, the amplification factor, is independent of the grid and plate
voltages. It is a measure of the relative effectiveness of grid and plate voltages in

producing electrostatic fields at the surfaces of the cathode. Placement of the control grid relative to the cathode and plate determines the amplification factor. The μ values of triodes generally range from 5 to 200. Key mathematical relationships include the following [6]:

$$0 = \frac{\hat{1}\,E_b}{\hat{1}\,E_{c1}}$$

$$R_p = \frac{\Delta E_b}{\Delta I_b}$$

$$S_m = \frac{\Delta I_b}{\Delta E_{c1}}$$

where:

    μ = amplification factor (with plate current held constant)

    R_p = dynamic plate resistance

    S_m = transconductance (also may be denoted G_m)

    E_b = total instantaneous plate voltage

    E_{c1} = total instantaneous control grid voltage

    I_b = total instantaneous plate current

The total cathode current of an ideal triode can be determined from

$$I_k = \left\{ E_c + \frac{E_b}{\mu} \right\}^{3/2}$$

where:

    I_k = cathode current

    K = a constant determined by tube dimensions

    E_c = grid voltage

    E_b = plate voltage

    μ = amplification factor

The grid of a triode usually consists of relatively fine wire wound on two supporting rods (*siderods*) and extending the length of the cathode [1]. The spacing between turns of wire is large compared with the size of the wire so that the passage of electrons from the cathode to the plate is essentially unobstructed by the grid. In some device types, a frame grid is used. The frame consists of two siderods supported by four metal straps. Fine lateral wire (diameter of 0.5 mil or less) is wound under tension around the frame. This type of grid permits the use of closer spacing between grid wires and between tube electrodes, and thus improves tube performance.

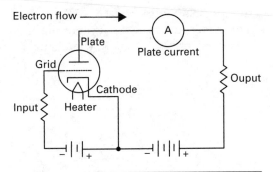

FIGURE 3.10 Basic triode circuit. (*After* [1].)

The purpose of the grid is to control the flow of plate current. When the tube is used as an amplifier, a negative DC voltage is usually applied to the grid. Under this condition the grid does not draw appreciable current. The number of electrons attracted to the plate depends on the combined effect of the grid and plate polarities, as shown in Figure 3.10. When the plate is positive, as is normal, and the DC grid voltage is made more negative, the plate is less able to attract electrons and the plate current decreases. When the grid is made less negative (more positive), the plate more readily attracts electrons to it and the plate current increases. Hence, when the voltage on the grid is varied in accordance with an input signal, the plate current varies with the signal. Because a small voltage applied to the grid can control a comparatively large amount of plate current, the signal is *amplified* by the tube.

The grid, plate, and cathode of a triode form an electrostatic system, each electrode acting as one plate of a small capacitor. The capacitances are those consisting between the grid and plate, plate and cathode, and grid and cathode. These capacitances are known as *interelectrode capacitances*.

Example Device

The 12AU7 is a twin triode used as a phase inverter, dual amplifier, push-pull amplifier, or oscillator in audio and radio equipment [1]. The miniature device utilizes a nine-pin socket. Each triode is independent of the other except for a common heater. The pin arrangement is shown in Figure 3.11. Operating characteristics are charted in Figure 3.12. Selected tube parameters are listed in Table 3.2.

Tetrode

The tetrode is a four-element tube with two grids [5]. The control grid serves the same purpose as the grid in a triode, while a second (*screen*) grid is mounted between the control grid and the anode. The elements of the screen grid are mounted directly behind the control-grid wires, as observed from the cathode surface, and serve as

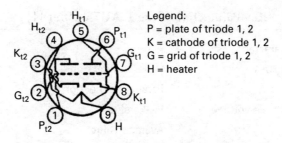

FIGURE 3.11   Pin arrangement for the 12AU7A/ECC82. (*After* [1].)

a shield or screen between the input circuit and the output circuit of the tetrode. The principal advantages of a tetrode over a triode include lower internal plate-to-grid feedback and lower drive power requirements.

Plate current is almost independent of plate voltage in a tetrode and therefore can be considered a *constant-current device.* The voltages on the screen and control grids determine the amount of plate current.

The total cathode current of an ideal tetrode is determined by

$$I_k = K \left\{ E_{c1} + \frac{E_{c2}}{\mu_s} + \frac{E_b}{\mu_p} \right\}^{\frac{3}{2}}$$

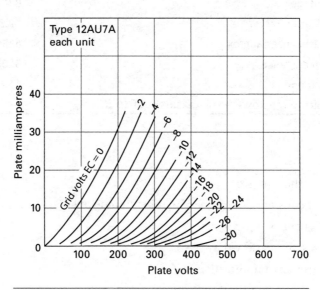

FIGURE 3.12   Operating characteristics of each section of the 12AU7A. (*After* [1].)

TABLE 3.2   Key Parameters of the 12AU7A (*After* [1].)

General Characteristics

Heater (AC/DC)	Series connection	12.6 V, 0.15 A	
	Parallel connection	6.3 V, 0.3 A	
Heater-cathode voltage	Peak value	±200 V, maximum	
	Average value	100 V maximum	
Direct interelectrode capacitance (approx.)	Grid to plate	1.5 pF, #1 and #2	
	Grid to cathode and heater	1.6 pF, #1 and #2	
	Plate to cathode and heater	0.5 pF #1, 0.35 pF #2	

Class A Amplifier (each unit, design maximum values)

Plate voltage		330 V
Cathode current		22 mA
Plate dissipation	Each plate	2.75 W
	Both plates (both units in operation)	5.5 W
Grid-circuit resistance	For fixed-bias operation	0.25 megohm
	For cathode-bias operation	1 megohm

Characteristics

Plate voltage	100 V	250 V
Grid voltage	0 V	−8.5 V
Amplification factor	19.5	17
Plate resistance (approx.)	6250 Ω	7700 Ω
Transconductance	3100 μmhos	2200 μmhos
Plate current	11.8 mA	10.5 mA
Grid voltage (approx.) for plate current of 10 μA	—	−24 V

where:

I_k = cathode current

K = a constant determined by tube dimensions

E_{c1} = control grid voltage

E_{c2} = screen grid voltage

μ_s = screen amplification factor

μ_p = plate amplification factor

E_b = plate voltage

The effectiveness of the shielding properties of the screen grid is increased through the use of a bypass capacitor connected between the screen grid and the cathode [1]. By means of the screen grid and the bypass capacitor, the grid-plate capacitance of a tetrode can be made quite small. In practice, the grid-plate capacitance of a receiving tube is reduced from several picofarads for a triode to 0.1 picofarad or less for a screen-grid tube.

The screen grid is operated at a positive voltage and, therefore, attracts electrons from the cathode. However, because of the comparatively large space between wires of the screen grid, most of the electrons drawn to the screen grid pass through it to the plate. Hence, the screen grid supplies an electrostatic force pulling electrons from the cathode to the plate. At the same time the screen grid shields the electrons between the cathode and screen grid from the plate so that the plate exerts very little electrostatic force on electrons near the cathode.

So long as the plate voltage is higher than the screen-grid voltage, plate current in a screen-grid tube depends to a great degree on the screen-grid voltage and very little on the plate voltage. The fact that plate current in a screen-grid tube is largely independent of plate voltage makes it possible to obtain much higher amplification with a tetrode than with a triode. The low grid-plate capacitance makes it possible to obtain this high amplification without plate-to-grid feedback and resultant instability. In receiving tube applications, the tetrode has been replaced in large part by the pentode.

The basic tetrode circuit is illustrated in Figure 3.13.

Pentode

In all electron tubes, electrons striking the plate may, if moving at sufficient speed, dislodge other electrons [1]. In two- and three-electrode types, these dislodged electrons usually do not cause problems because no positive electrode other than the plate itself is present to attract them. These electrons, therefore, are drawn back to the plate. Emission caused by bombardment of an electrode by electrons from the cathode is called secondary emission.

In the case of screen-grid tubes, the proximity of the positive screen grid to the plate offers a strong attraction to these secondary electrons, and particularly so if the plate voltage swings lower than the screen-grid voltage. This effect reduces the plate current and limits the useful plate voltage swing for a tetrode.

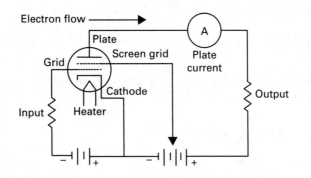

FIGURE 3.13   Tetrode operating circuit. (*After* [1].)

The pentode is a five-electrode tube incorporating three grids [5]. The control and screen grids perform the same function as in a tetrode. The third grid, the *suppressor grid*, is mounted in the region between the screen grid and the anode. The suppressor grid produces *a potential minimum*, which prevents secondary electrons from being interchanged between the screen and plate. The pentode's main advantages over the tetrode include the following:

- Reduced secondary emission effects.
- Good linearity.
- Ability to let plate voltage swing below the screen voltage without excessive screen dissipation. This allows slightly higher power output for a given operating plate voltage.

Because of the design of the pentode, plate voltage has even less effect on plate current than in the tetrode. The same total space-current equation applies to the pentode as with the tetrode:

$$I_k = K \left\{ E_{c1} + \frac{E_{c2}}{\mu_s} + \frac{E_b}{\mu_p} \right\}^{3/2}$$

where:

I_k = cathode current

K = a constant determined by tube dimensions

E_{c1} = control grid voltage

E_{c2} = screen grid voltage

μ_s = screen amplification factor

μ_p = plate amplification factor

E_b = plate voltage

The suppressor grid may be operated negative or positive with respect to the cathode. It also may be operated at cathode potential. It is possible to control plate current by varying the potential on the suppressor grid. The basic pentode circuit is illustrated in Figure 3.14.

Example Device

The 6K6 power pentode is a glass octal type used in the output stage of amplifiers of various types [1]. The pin configuration is shown in Figure 3.15. Selected tube parameters are listed in Table 3.3.

Beam Power Tube

A beam power tube is a tetrode or pentode in which directed electron beams are used to increase substantially the power-handling capability of the device [1]. Such a tube contains a cathode, control grid (grid #1), screen grid (grid #2), plate, and optionally

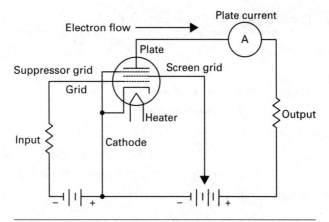

FIGURE 3.14   Basic pentode circuit. (*After* [1].)

a suppressor grid (grid #3). When a beam power tube is designed without a suppressor grid, the electrodes in the device are so spaced that secondary emission from the plate is suppressed by space-charge effects between the screen grid and the plate. The space charge is produced by the slowing of electrons traveling from a high-potential screen grid to a lower-potential plate. In this low-velocity region, the space charge produced is sufficient to repel secondary electrons emitted from the plate and to cause them to return to the plate.

Beam power tubes of this design employ beam-confining electrodes at cathode potential to assist in producing the desired beam effects and to prevent stray electrons from the plate from returning to the screen grid outside the beam. One useful feature of a beam power tube is its low screen-grid current. The screen grid and the control grid are spiral wires wound so that each turn of the screen grid is shaded from the cathode by a grid turn. This alignment of the screen grid and control grid causes the electrons to travel in "sheets" between the turns of the screen grid so that very few of them strike the screen grid. Because of the effective suppressor action provided

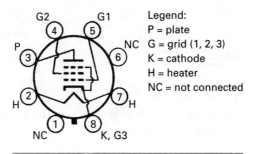

FIGURE 3.15   Connection diagram of the 6K6GT. (*After* [1].)

TABLE 3.3   Key Parameters for the 6K6GT (*After* [1].)

General Characteristics

Heater (AC/DC)	Series connection	6.3 V, 0.4 A
Heater-cathode voltage	Peak value	±200 V
	Average value	100 V
Direct interelectrode capacitance (approx.)	Grid #1 to plate	0.5 pF
	Grid #1 to cathode, heater, grid #2, and grid #3	5.5 pF
	Plate to cathode, heater, grid #2, and grid #3	6 pF

Class A Amplifier (maximum ratings, design center values)

Plate voltage		315 V
Grid #2 (screen grid) voltage		285 V
Plate dissipation		8.5 W
Grid #2 input		2.8 W
Grid #1 circuit resistance	For fixed-bias operation	0.1 megohm
	For cathode-bias operation	0.5 megohm

Class A Amplifier (typical operation)

Plate voltage	100 V	250 V	315 V
Grid #2 voltage	100 V	250 V	250 V
Grid #1 (control grid) voltage	–7 V	–18 V	–21 V
Peak AF grid #1 (control grid) voltage	7 V	18 V	21 V
Zero signal plate current	9 mA	32 mA	25.5 mA
Maximum-signal plate current	9.5 mA	33 mA	28 mA
Zero-signal grid #2 current	1.6 mA	5.5 mA	4.0 mA
Maximum-signal grid #2 current	3 mA	10 mA	9 mA
Plate resistance (approx.)	104000 Ω	90000 Ω	110000 Ω
Transconductance	1500 µmhos	2300 µmhos	2100 µmhos
Load resistance	12000 Ω	7600 Ω	9000 Ω
Maximum-signal power output	0.35 W	3.4 W	4.5 W

by space charge, and because of the low current drawn by the screen grid, the beam power tube has the advantage of high power output, high power sensitivity, and high efficiency.

Figure 3.16 shows the structure of a beam power tube employing space-charge suppression and illustrates how the electrons are confined. The beam condition illustrated is that for a plate potential less than the screen-grid potential.

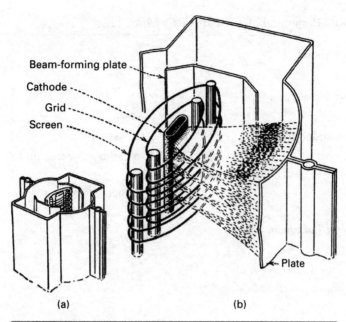

FIGURE 3.16 Structure of the beam power tube showing the beam-confining action of the elements: (*a*) external view, (*b*) cutaway view. (*After* [5].)

Example Device

The 6L6 is used in the output stage of audio amplifying equipment, especially units designed for high power output capability [1]. The pin arrangement for this octal socket device is shown in Figure 3.17. Characteristics of the 6L6 are charted in Figure 3.18. Selected tube parameters are listed in Table 3.4.

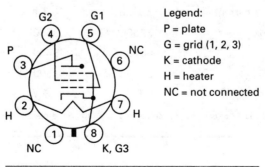

FIGURE 3.17 Pin arrangement of the 6L6GC. (*After* [1].)

TABLE 3.4   Key Parameters for the 6L6GC. (*After* [1].)

General Characteristics

Heater (AC/DC)		6.3 V, 0.9 A
Heater-cathode voltage	Peak value	±200 V
	Average value	100 V
Direct interelectrode capacitance (approx.)	Grid #1 to plate	0.6 pF
	Grid #1 to cathode, heater, grid #2, and grid #3	10 pF
	Plate to cathode, heater, grid #2, and grid #3	65 pF

Class A Amplifier (maximum ratings)

Plate voltage		500 V
Grid #2 (screen grid) voltage		450 V[1]
Plate dissipation		30 W
Grid #2 input		5 W
Grid #1 circuit resistance	For fixed-bias operation	0.1 megohm
	For cathode-bias operation	0.5 megohm

Class A Amplifier (typical operation)

Plate voltage	250 V	300 V	350 V
Grid #2 voltage	250 V	200 V	250 V
Grid #1 (control grid) voltage	−14 V	−12.5 V	−18 V
Peak AF grid #1 (control grid) voltage	14 V	12.5 V	18 V
Zero signal plate current	72 mA	48 mA	54 mA
Maximum-signal plate current	79 mA	55 mA	66 mA
Zero-signal grid #2 current	5 mA	2.5 mA	2.5 mA
Maximum-signal grid #2 current	7.3 mA	4.7 mA	7 mA
Plate resistance (approx.)	225000 Ω	35000 Ω	33000 Ω
Transconductance	6000 μmhos	5300 μmhos	5200 μmhos
Load resistance	2500 Ω	4500 Ω	4200 Ω
Maximum-signal power output	6.5 W	6.5 W	10.8 W

[1] In push-pull circuits where grid #2 of each tube is connected to a tap on the plate winding of the output transformer, this maximum rating is 500 V.

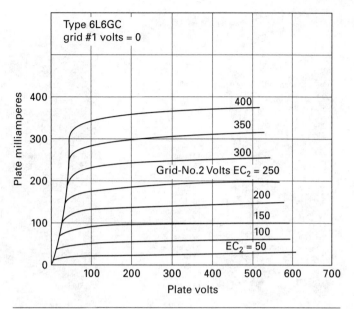

FIGURE 3.18   Operating characteristics of the 6L6GC. (*After* [1].)

Vacuum Tube Structure

Receiving tubes generally utilize a glass envelope and a base of molded phenolic material or a glass base with connecting pins extending outside the vacuum envelope. The most common pinouts include octal (eight-pin), and miniature seven-pin and nine-pin devices. A detailed diagram of an octal base tube is shown in Figure 3.19*a*. A cutaway view of a miniature tube is illustrated in Figure 3.19*b*. Additional styles have been produced and can be found in some equipment.

Vacuum Tube Design

Any particular vacuum tube may be designed to meet a number of operating parameters, the most important usually being high operating efficiency and high gain/bandwidth properties [5]. Above all, the tube must be reliable and provide long operating life. The design engineer must examine a laundry list of items, including:

- **Cooling**   How the tube will dissipate heat generated during normal operation. A high-performance tube is of little value if it will not provide long life in typical applications.
- **Electro-optics**   How the internal elements line up to achieve the desired performance. A careful analysis must be made of what happens to the electrons in their paths from the cathode to the anode, including the expected power gain of the tube.

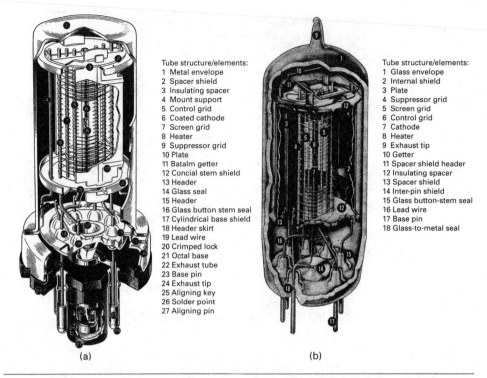

Tube structure/elements:
1 Metal envelope
2 Spacer shield
3 Insulating spacer
4 Mount support
5 Control grid
6 Coated cathode
7 Screen grid
8 Heater
9 Suppressor grid
10 Plate
11 Batalm getter
12 Concial stem shield
13 Header
14 Glass seal
15 Header
16 Glass button stem seal
17 Cylindrical base shield
18 Header skirt
19 Lead wire
20 Crimped lock
21 Octal base
22 Exhaust tube
23 Base pin
24 Exhaust tip
25 Aligning key
26 Solder point
27 Aligning pin

Tube structure/elements:
1 Glass envelope
2 Internal shield
3 Plate
4 Suppressor grid
5 Screen grid
6 Control grid
7 Cathode
8 Heater
9 Exhaust tip
10 Getter
11 Spacer shield header
12 Insulating spacer
13 Spacer shield
14 Inter-pin shield
15 Glass button-stem seal
16 Lead wire
17 Base pin
18 Glass-to-metal seal

(a)   (b)

FIGURE 3.19   Cutaway views of vacuum tubes: (*a*) metal envelope pentode, (*b*) miniature glass envelope type. (*After* [5] *and* [1], *respectively.*)

- **Operational parameters**   What the typical interelectrode capacitances will be and the manufacturing tolerances that can be expected. This analysis includes spacing variations among elements within the tube, the types of materials used in construction, the long-term stability of the internal elements, and the effects of thermal cycling.

Tube Elements

The ultimate performance of any vacuum tube is determined by the accuracy of design and construction of the internal elements. The cathode obtains the energy required for electron emission from heat.

Heater/Cathode

The cathode may be directly heated or indirectly heated, as illustrated in Figure 3.20 [1]. A directly heated cathode, or filament-cathode, is a wire heated by the passage of an electric current. An indirectly heated cathode, or heather-cathode, consists of a filament (heater) enclosed in a metal sleeve. The sleeve carries the electron-emitting

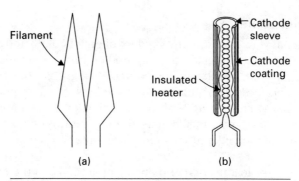

FIGURE 3.20   Tube structures: (*a*) filament or directly heated cathode, (*b*) indirectly heated cathode, or heather-cathode. (*After* [1].)

material on its outside surface and is heated by radiation and conduction from the heater element.

Among the common types of emitting surfaces are

- **Alkaline-earth oxides**   Alkaline earths are usually applied as a coating on a nickel-alloy wire or ribbon. This coating, which is dried in a relatively thick layer on the filament, requires a comparatively low temperature of about 700 to 750°C to produce a sufficient supply of electrons for a given tube. Coated filaments operate quite efficiently and require relatively little filament power.
- **Tungsten**   Tungsten filaments are made from the pure metal and must operate at very high temperatures to achieve sufficient electron emission. Tungsten filaments thus require a relatively large amount of filament power compared to other filament types.
- **Thoriated-tungsten**   Thoriated-tungsten filaments are made from tungsten impregnated with thorium oxide. Due to the presence of thorium, these filaments liberate electrons at a temperature of about 1700°C and are, therefore, more economical on filament power than pure tungsten filaments.

From the standpoint of circuit design, the heater-cathode construction offers advantages in connection flexibility because of the electrical separation of the heater from the cathode. In addition, the cathode sleeve serves to shield other elements of the tube from the introduction of hum from the AC heater power supply.

Another advantage of the heater-cathode construction is that it makes practical the design of a rectifier tube having close spacing between the cathode and plate, and of an amplifier tube having close spacing between the cathode and control grid. In a close-spaced rectifier tube, the voltage drop in the device is low and, therefore, regulation is improved. In an amplifier tube, close spacing increases the gain obtainable from the device. Because of these and other advantages, almost all receiving tubes utilize heather-cathode construction.

Grid Elements

Conventional wire grids are prepared by operators that wind the assemblies using *mandrels* (forms) that include the required outline of the finished grid. The operators spot-weld the wires at intersecting points. Most grids of this type are made with tungsten or *molybdenum,* which exhibit stable physical properties at elevated temperatures. On a strength basis, pure molybdenum generally is considered the most suitable of all refractory metals at temperatures of 870 to 1650°C. The thermal conductivity of molybdenum is more than three times that of iron and almost half that of copper.

The external loading of a grid during operation and the proximity of the grid to the hot cathode impose considerable demands on both the mechanical stability of the structure and the physical characteristics of its surface. The grid absorbs a high proportion of the heat radiated by the cathode. It also intercepts the electron beam, converting part of its kinetic energy into heat. The result is that grids are forced to work in high temperatures. Their primary and secondary emissions, however, must be low. To prevent grid emission, high electron affinity must be ensured throughout the life of the tube, even though it is impossible to prevent material that evaporates from the cathode from contaminating the grid surface to some extent.

High-Frequency Operating Limits

As with most active devices, performance of a given vacuum tube deteriorates as the operating frequency is increased beyond its designed limit [5]. Electron *transit time* is a significant factor in the upper-frequency limitation of electron tubes. A finite time is taken by electrons to traverse the space from the cathode, through the grid(s), and travel on to the plate. As the operating frequency increases, a point is reached at which the electron transit-time effects become significant. This point is a function of the accelerating voltages at the grid(s) and the anode and their respective spacing. Tubes with reduced spacing in the grid-to-cathode region exhibit reduced transit-time effects. Transit time typically is not a problem for tubes operating below 30 MHz. A power limitation also is interrelated with the high-frequency limit of a device. As the operating frequency is increased, closer spacing and smaller-sized electrodes must be used. This reduces the power-handling capability of the tube.

Tube Assembly

Each type of vacuum tube is unique insofar as its operating characteristics are concerned [2]. The basic physical construction, however, is common to most devices. After the components have been assembled and mounted inside the glass envelope, the device goes through a *bake-out* procedure. Baking stations are used to evacuate the tube and bake out any oxygen or other gases from parts of the assembly to ensure long component life.

A vacuum offers excellent electrical insulation characteristics. This property is essential for reliable operation of a vacuum tube, the elements of which typically operate at high potentials with respect to each other and to the surrounding environment. An electrode containing absorbed gases, however, will exhibit reduced breakdown voltage because the gas will form on the electrode surface, increasing the surface gas pressure and lowering the breakdown voltage in the vicinity of the gas pocket.

To maintain a high vacuum during the life of the component, certain tubes contain a *getter* device. The name comes from the function of the element: to "get," or trap and hold, gases that may exist inside the tube. Materials used for getters include zirconium, cerium, barium, and titanium.

The operation of a vacuum tube is an evolving chemical process. End of life in a vacuum tube generally is caused by loss of emission. Tube failure can also occur as a result of a broken filament or a short-circuit between tube elements caused by physical stress.

Neutralization

An amplifier operating at high frequencies must be properly neutralized to provide acceptable performance in most applications [5]. The means to accomplish this end vary considerably from one design to another. An amplifier is neutralized when two operating conditions are met:

- The interelectrode capacitance between the input and output circuits is canceled.
- The inductance of the screen grid and cathode assemblies (in a tetrode) is canceled.

Cancellation of these common forms of coupling between the input and output circuits of vacuum tube amplifiers prevents self-oscillation and the generation of spurious products.

Figure 3.21 illustrates the primary elements that affect neutralization of a vacuum tube amplifier operating in the very high frequency (VHF) band. (Many of the following principles also apply to lower frequencies.) The feedback elements include the residual grid-to-plate capacitance (C_{gp}), plate-to-screen capacitance (C_{ps}), and screen-grid lead inductance (L). The radio frequency (RF) energy developed in the plate circuit (E_p) causes a current (I) to flow through the plate-to-screen capacitance and the screen lead inductance. The current through the screen inductance develops a voltage ($-E$) with a polarity opposite that of the plate voltage (E_p). The $-E$ potential often is used as a method of neutralizing tetrode and pentode tubes operating in the VHF band.

Figure 3.22 graphically illustrates the electrical properties at work. The circuit elements of the previous figure have been arranged so that the height above or below the zero potential line represents the magnitude and polarity of the RF voltage for each part of the circuit with respect to ground (zero). For the purposes of this illustration, assume that all of the circuit elements involved are pure reactances. The voltages represented by each, therefore, are either in phase or out of phase and can be treated as positive or negative with respect to each other.

The voltages plotted in the figure represent those generated as a result of the RF output circuit voltage (E_p). No attempt is made to illustrate the typical driving

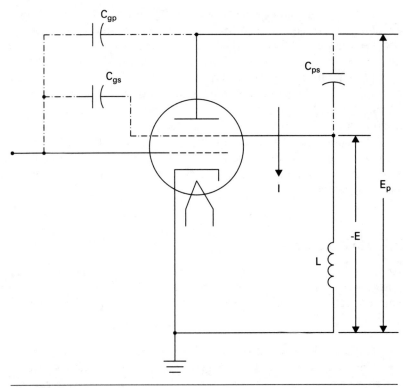

FIGURE 3.21 The elements involved in the neutralization of a tetrode PA stage. (*After* [5].)

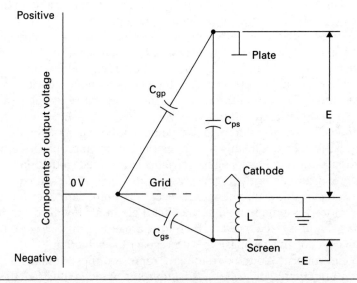

FIGURE 3.22 A graphical representation of the elements involved in the self-neutralization of a tetrode RF stage. (*After* [5].)

current on the grid of the tube. The plate (P) has a high positive potential above the zero line, established at the ground point. Keep in mind that the distance above the baseline represents increasing positive potential. The effect of the out-of-phase screen potential developed as a result of inductance L_s is shown, resulting in the generation of $-E$.

As depicted, the figure constitutes a perfectly neutralized circuit. The grid potential rests at the zero baseline. The grid operates at filament potential insofar as any action of the output circuit on the input circuit is concerned.

The total RF voltage between plate and screen is made up of the plate potential and screen lead inductance voltage, $-E$. This total voltage is applied across a divider circuit that consists of the grid-to-plate capacitance and grid-to-screen capacitance (C_{gp} and C_{gs}). When this potential divider is properly matched for the values of plate RF voltage (Ep) and screen lead inductance voltage ($-E$), the control grid will exhibit zero voltage difference with respect to the filament as a result of E_p.

A variety of methods may be used to neutralize a vacuum tube amplifier. Generally speaking, a grounded-grid, cathode-driven triode can be operated into the VHF band without external neutralization components. The grounded-grid element is sufficient to prevent spurious oscillations. Tetrode amplifiers generally will operate through the Medium Frequency (MF) band without neutralization. However, as the gain of the stage increases, the need to cancel feedback voltages caused by tube interelectrode capacitances and external connection inductances becomes more important.

For operation at frequencies below the VHF band, neutralization for a tetrode typically employs a capacitance bridge circuit to balance out the RF feedback caused by residual plate-to-grid capacitance. This method assumes that the screen is well bypassed to ground, providing the expected screening action inside the tube.

Electron Tube Characteristics

The term "characteristics" is used to identify the distinguishing electrical features and values of an electron tube [1]. These values may be shown in curve form or they may be tabulated (or both). When the characteristic values are given in curve form, the curves may be used to determine tube performance and to calculate additional factors. Tube characteristics are obtained from electrical measurements of a device in various circuits under defined conditions.

Static characteristics may be shown by plate characteristics curves and transfer (mutual) characteristics curves. These curves present the same information but in two different forms to increase their usefulness:

- Plate characteristics curves are obtained by varying the plate voltage and measuring plate current for different grid-bias voltages.
- Transfer characteristics curves are obtained by varying the grid-bias voltage and measuring plate current for different plate voltages.

A plate characteristics family of curves is shown in Figure 3.23, while Figure 3.24 illustrates the transfer characteristics family of curves for the same tube.

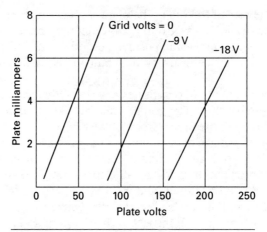

FIGURE 3.23   Family of plate characteristics curves. (*After* [1].)

Dynamic characteristics include amplification factor, plate resistance, control-grid plate transconductance, plate efficiency, power sensitivity, and other parameters. These characteristics may be shown in curve form for variations in specified operating conditions. Specifically:

- **Amplification factor** (μ)   The ratio of the change in plate voltage to a change in control-electrode voltage in the opposite direction, under the conditions that the plate current remains unchanged and that all other electrode voltages are maintained constant. The μ of a device is useful for calculating stage gain.

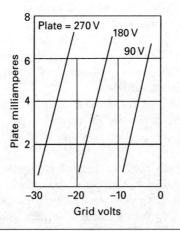

FIGURE 3.24   Family of transfer characteristics curves. (*After* [1].)

Example: If, when the plate voltage is made 1 volt more positive, the control-electrode (grid #1) voltage must be made 0.1 volt more negative to hold plate current unchanged, then the amplification factor is equal to 1 divided by 0.1, or 10. In other words, a small voltage variation in the grid current of the tube has the same effect on plate current as a large plate voltage change—the latter equal to the product of the grid voltage change and the amplification factor.

- **Plate resistance (r_p)**   The resistance of the path between the cathode and plate to the flow of alternating current. This parameter is the quotient of a small change in plate voltage divided by the corresponding change in plate current, and is expressed in ohms.
  Example: If a change of 0.1 milliampere (0.0001 ampere) is produced by a plate voltage variation of 1 volt, then the plate resistance is equal to 1 divided by 0.0001, or 100,0000 ohms.

- **Transconductance (g_m)**   A factor that combines in one term the amplification factor and the plate resistance, and is the quotient of the first divided by the second. More correctly described as control grid-to-plate transconductance, this term also has been known as *mutual conductance*. Transconductance may be more strictly defined as the quotient of a small change in plate current (amperes) divided by the small change in control-grid voltage producing it, under the condition that all other voltages remain unchanged.
  Example: If a grid voltage change of 0.5 volts causes a plate current change of 1 milliampere (0.001 ampere), with all other voltages held constant, then the transconductance is equal to 0.001 divided by 0.5, or 0.005 mho ("mho" is the unit of conductance). For convenience, a millionth of a mho, or a micromho (μmho), is commonly used to express transconductance. Thus, in this example, 0.002 mho is equal to 2000 micromhos.

- **Plate efficiency**   The ratio of the AC power output (P_o) of an amplifier tube to the product of the average DC plate voltage (E_b) and the DC plate current (I_b) at full signal level:

$$\text{Plate efficiency (\%)} = \frac{P_o}{E_b \times I_b} \times 100$$

  where P_o = AC power output in watts, E_b = average DC plate voltage in volts, and I_b = DC plate current in amperes.

- **Power sensitivity**   The ratio of the power output (P_o) of an amplifier tube to the square of the input signal voltage (E_{in}), expressed in mhos:

$$\text{Power senstivity (mhos)} = \frac{P_o}{E_{in}^2}$$

  where P_o = AC power output in watts and E_{in} = input signal voltage (rms).

Interpretation of Tube Data

The data provided by electron tube manufacturers includes typical operating values, characteristics, and characteristic curves [1]. Ratings are established on electron tube types to help equipment designers utilize the performance and service capabilities of each tube type to the best advantage. Ratings are given for those characteristics that careful study and experience indicate must be kept within certain limits to ensure satisfactory performance.

Various rating systems have been used by the electron tube industry over the years, including the following:

- **Absolute Maximum**   Limiting values that should not be exceeded with any tube of the specified type under any condition of operation. These ratings are not used too often for receiving tubes, but are generally used for transmitting and industrial tubes.
- **Design Center**   Limiting values that should not be exceeded with a tube of the specified type having characteristics equal to the published values under normal operating conditions. These ratings, which include allowances for normal variations in both tube characteristics and operating conditions, were used for most receiving tubes prior to 1957.
- **Design Maximum**   Limiting values that should not be exceeded with a tube of the specified type having limiting characteristics equal to the published values under any conditions of operation. These ratings include allowances for normal variations in tube characteristics, but do not provide for variations in operating conditions. Design Maximum ratings were adopted for receiving tubes in 1957.

Many ratings, such as voltage and current ratings, are in general self-explanatory, but additional detail is useful:

- **Plate Dissipation**   The power dissipated in the form of heat by the plate as a result of electron bombardment. This value is the difference between the power supplied to the plate of the tube and the power delivered by the tube to the load.
- **Peak Heater-Cathode Voltage**   The highest instantaneous value of voltage that a tube can safely stand between its heater and cathode. This rating is applied to tubes having a separate cathode terminal, and is used in applications where excessive voltage may be introduced between the heater and the cathode.
- **Maximum DC Output Current**   The highest average plate current that can be handled continuously by a rectifier tube. This value for any rectifier is based on the permissible plate dissipation of that tube type. Under operating conditions involving a rapidly repeating duty cycle (steady load), the average plate current may be measured with a DC ammeter.
- **Maximum Peak Plate Current**   The highest instantaneous plate current that a tube can safely carry recurrently in the direction of normal current flow.

- **Maximum Peak Inverse Plate Voltage**   The highest instantaneous plate voltage that the tube can withstand recurrently in the direction opposite to that in which it is designed to pass current.
- **Typical Operation Values**   Values for typical operation given to serve as guiding information for the use of each tube type. These values should not be confused with *ratings,* because a tube can be used under any suitable conditions within its maximum ratings, according to the application. The power output value for any operating condition is an approximate tube output; that is, plate input minus plate loss. Circuit losses must be subtracted from the tube output in order to determine the useful output.
- **Interelectrode Capacitance**   Discrete capacitances measured between specified elements or groups of elements in electron tubes. Unless otherwise indicated, all capacitances are measured with filament or heater cold, with no direct voltages present, and with no external shields installed.
- **Grid #2 (screen grid) Input**   The power applied to the grid #2 electrode, which consists essentially of the power dissipated in the form of heat by the grid as a result of electron bombardment.

Tube Pinout

Terminal diagrams are given in the technical data provided by tube manufacturers. Terminal numbering is usually clockwise as viewed from the bottom of the device [1]. Some common pinouts are illustrated in Figure 3.25. For nearly all modern tubes, the spacing between pin #1 and the highest number pin is somewhat greater than the spacing between all of the other pins. For octal-based types, the key for orienting the tube when it is inserted in a socket also serves to designate the #1 pin, which is the first pin clockwise from the key.

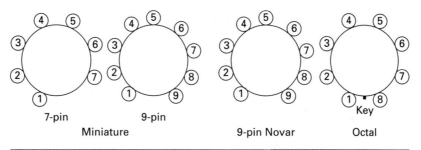

FIGURE 3.25   Terminal connections for common tube types (bottom view). (*After* [1].)

Chapter 4

Vacuum Tube Circuits

Vacuum tubes have been used for decades to perform an amazing array of functions—from radio to radar and beyond. This chapter will focus on audio applications and their supporting circuits. The performance characteristics of an audio amplifier are given in terms of frequency response, distortion, noise, and power output. The maximum power output that a high-fidelity amplifier can deliver depends upon a complex relation of factors and impacts other characteristics of the amplifier.

It is difficult to separate the amplifier from the environment in which it will operate. The interface with the listening environment is—invariably—a loudspeaker. The acoustic power required to reproduce the loudest passages of orchestral music at concert-hall level in an average size living room has been estimated to be about 0.4 W.[1] For the sake of argument, we will round it up to 1 W. It is important to note that loudspeakers are quite inefficient from a power standpoint. Only a small percentage of the electrical energy sent by the amplifier to the speaker is actually converted to acoustic energy; estimates range from 1 percent to 5 percent or so (the remaining energy is converted to heat). This being the case, an amplifier would need to deliver 20 W (for the best-case loudspeaker efficiency of 5 percent) to 100 W (for the worst-case loudspeaker efficiency of 1 percent) to the load in order to satisfy the needs of the consumer. The power required by an amplifier dictates to a considerable extent the basic design choices.

Amplifier Types

Amplifiers are classified in ways descriptive of their characteristics and properties. A first-level sort leads to four broad classes based on the frequencies to be amplified:

- Direct-current amplifier
- Audio frequency amplifier
- Video amplifier
- Radio frequency amplifier

---

[1] See [2], page 49. The term "average-size" living room is admittedly vague.

For the purposes of this chapter, we will focus on audio frequency amplifiers, which can be generally described to cover a range of 20 Hz to 20 kHz.

Amplifiers also can be divided into voltage and power amplifiers according to whether the objective is to produce as much voltage or as much power as possible in the load impedance. The output stage of an audio amplifier is typically a power amplifier, while the preceding stages are typically voltage amplifiers.

The type of amplifier is fixed primarily by the constants of the associated electrical circuits and by the grid bias and plate voltages employed. It is possible to make most any particular tube function as any kind of amplifier, although the tube characteristics best suited for each type of amplifier vary, sometimes considerably. These differences have given rise to tubes designed to function best in a particular type of service.

A number of different types of amplifiers are possible and practical for audio use, each offering a particular set of features, attributes, and constraints. Circuit design is an exercise in compromise, typically trading off a number of variables that includes—but is not limited to—cost, complexity, performance, power consumption, and reliability.

Operating Class

An electron-amplifying device is classified by its individual class of operation [1]. Three primary class divisions apply to vacuum tube devices:

- **Class A** A mode wherein the power-amplifying device is operated over its linear transfer characteristic. This mode provides the lowest waveform distortion, but also the lowest efficiency. The basic operating efficiency of a class A stage is 50 percent. Class A amplifiers exhibit low distortion, making them well suited to audio amplifier applications.
- **Class B** A mode wherein the power-amplifying device is operated just outside its linear transfer characteristic. This mode provides improved efficiency at the expense of some waveform distortion. Class AB is a variation on class B operation. The transfer characteristic for an amplifying device operating in this mode is, predictably, between class A and class B.
- **Class C** A mode wherein the power-amplifying device is operated significantly outside its linear transfer characteristic, resulting in a pulsed output waveform. High efficiency (up to 90 percent) can be realized with class C operation, but significant distortion of the waveform will occur. Class C is used extensively as an efficient radio frequency (RF) power generator.

The operating class also can be viewed as a function of the angle of current flow. Typically, the following generalizations regarding conduction angle apply [2]:

- **Class A** 360°. The grid bias and alternating grid voltages are such that plate current in a specific tube flows at all times.
- **Class AB** Between 180 and 360°. The grid bias and alternating grid voltages are such that plate current in a specific tube flows for appreciably more than half but less than the entire electrical cycle.

- **Class B**   180°. The grid bias is approximately equal to the cutoff value, so that: 1) plate current in a specific tube is approximately zero when no exciting grid voltage is applied, and 2) plate current in the tube flows for approximately one-half of each cycle when an alternating grid voltage is applied.
- **Class C**   Less than 180°. The grid bias is appreciably greater than the cutoff value, so that: 1) plate current in a specific tube is zero when no alternating grid voltage is applied, and 2) plate current in the tube flows for appreciably less than one-half of each cycle when an alternating grid voltage is applied.

Subscripts also may be used to denote grid current flow. The subscript "1" means that no grid current flows in the stage; the subscript "2" denotes grid current flow.

Class A amplifiers are used in applications requiring low harmonic distortion in the output signal. Typical operating plate efficiency for a class A amplifier is about 30 percent. Power gain is high because of the low drive power required. Gains as high as 30 dB are typical.

For audio frequency (AF) amplifiers in which distortion is an important factor, only class A amplification permits single-tube operation. In this case, operating conditions are usually chosen so that distortion is maintained at a low level, often through the use of *inverse feedback*. Reduced distortion with improved power performance also can be obtained by using a push-pull stage. With class AB and class B amplifiers, a balanced arrangement using two tubes is required for audio service.

A class AB power amplifier is capable of generating more power—using the same tube—than the class A amplifier, but more distortion also will be generated. A class B RF linear amplifier will generate still more distortion, but is acceptable in certain RF applications. The operating plate efficiency is typically 66 percent, and stage gain is about 20 to 25 dB. Class B operation is commonly used in push-pull configurations as a way of generating high power.

A class C power amplifier is used where large amounts of RF energy need to be generated with high efficiency. Class C RF amplifiers must be used in conjunction with tuned circuits or cavities, which restore the amplified waveform through the *flywheel effect*.

Class A Voltage Amplifier

In a class A voltage amplifier, the electron tube is used to reproduce grid voltage variations across an impedance or a resistance in the plate circuit [2]. These variations are essentially of the same form as the input signal voltage impressed on the grid, but their amplitude is increased. This increase is accomplished by operating the tube at a suitable grid bias so that the applied grid input voltage produces plate current variations proportional to the input signal swings. Because the voltage variation obtained in the plate circuit is much larger than that required to swing the grid, *amplification* of the signal is obtained. Figure 4.1 provides a graphical illustration of this process.

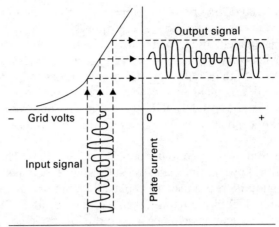

FIGURE 4.1   Current characteristics of a class A amplifier. (*After* [2].)

The plate current flowing through the load resistance causes a voltage drop that varies directly with the plate current. (See Figure 4.2.) The ratio of this voltage variation produced in the load resistance to the input signal voltage is the *voltage amplification,* or *gain,* provided by the tube. The voltage amplification due to the tube is expressed by the formula

$$\text{Voltage amplification} = \frac{\mu \times R_L}{R_L + r_p}$$

or

$$\frac{g_m \times r_p \times R_L}{1000000 \times (r_p + R_L)}$$

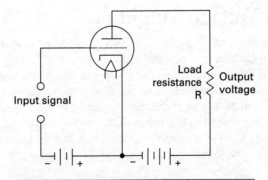

FIGURE 4.2   Triode amplifier circuit. (*After* [2].)

where:

μ = amplification factor of the tube

R_L = load resistance in ohms

r_p = plate resistance in ohms

g_m = transconductance in micromhos

From the first formula, it can be seen that the gain actually obtainable from the tube is less than the tube amplification factor, but that the gain approaches the amplification factor when the load resistance is large compared to the tube-plate resistance. This characteristic is charted in Figure 4.3. From the curve it can be seen that a high value of load resistance should be used to obtain high gain in a voltage amplifier. However, the plate resistor should not be too large, because the flow of plate current through the plate resistor produces a voltage drop that reduces the plate voltage applied to the tube.

The input impedance of a tube (i.e., the impedance between the grid and cathode) consists of the following component elements:

- A reactive component due to the capacitance between the grid and cathode
- A resistive component resulting from the transit time of electrons between the cathode and grid
- A resistive component developed by the part of the cathode lead inductance that is in common to both the input and output circuits

These components are dependent on the frequency of the applied signal. The input impedance is very high at audio frequencies when the tube is operated with its grid biased negative. In a class A_1 or AB_1 transformer-coupled audio amplifier,

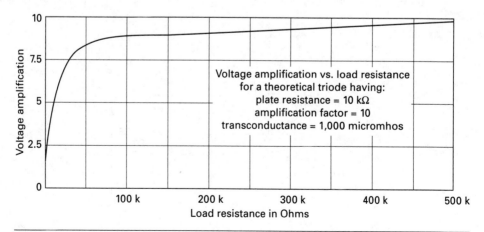

FIGURE 4.3   Gain curve for a triode amplifier circuit. (*After* [2].)

therefore, the loading imposed by the grid on the input transformer is negligible. As a result, the secondary impedance of a class A_1 or AB_1 input transformer can be made very high because the choice is not limited by the input impedance of the tube. Transformer design considerations may be the limiting factor.

At higher radio frequencies, the input impedance may become very low even when the grid is negative, due to the finite time of passage of electrons between the cathode and grid, and the appreciable lead reactance. This impedance drops rapidly as the frequency is raised and increases input-circuit loading.

Class A Power Amplifier

A power amplifier is used in applications where high power output is of more importance than high voltage amplification [2]. Gain possibilities, therefore, are traded in the design of a power tube to obtain greater power-handling capability. Triodes, pentodes, and beam power tubes designed for power amplifier service have certain inherent features; in general:

- Triode power tubes are characterized by low power sensitivity, low plate power efficiency, and low distortion.
- Pentode power tubes are characterized by high power sensitivity, high plate power efficiency, and usually somewhat higher distortion than class A triodes.
- Beam power tubes have higher power sensitivity and efficiency than triode or conventional pentode types.

A class A power amplifier can be used as a driver stage to supply power to a class AB_2 or class B stage. It is usually advisable to use a triode rather than a pentode in a driver stage because of the lower plate impedance of the triode.

Power tubes connected in either parallel or push-pull configurations may be employed as class A amplifiers to obtain increased output. The parallel connection (shown in Figure 4.4) provides twice the output of a single tube with the same value

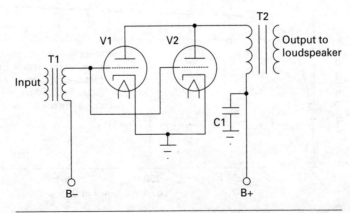

FIGURE 4.4   Power amplifier with tubes connected in parallel. (*After* [2].)

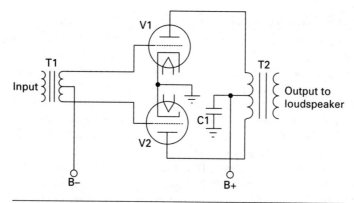

FIGURE 4.5   Power amplifier with tubes connected in push-pull. (*After* [2].)

of grid signal voltage. With this connection, the effective transconductance of the stage is doubled, and the effective plate resistance and the load resistance required are reduced by half as compared with a single tube.

The push-pull connection (shown in Figure 4.5) requires twice the grid signal voltage but provides increased output power and has other important advantages over single tube operation. Among the advantages (assuming identical tubes) are [4]:

- No direct-current saturation in the core of the output transformer. The direct currents in the two halves of the primary magnetize the core in opposite directions, and so produce zero resultant magnetism.
- There is no current of signal frequency flowing through the source of plate power. This means the push-pull power amplifier produces no regeneration event when there is a plate impedance common to the power and other stages; it also means that no bypass capacitor is required across the cathode-biasing resistor (if used).
- Alternating-current hum voltages present in the source of plate power produce no hum in the output because the hum currents flowing in the two halves of the primary balance each other.[2]
- There is less distortion for the same power output per tube, or more power output per tube for the same distortion, as a result of cancellation of even harmonics and even-order combination frequencies.

For either parallel or push-pull class A operation of two tubes, all electrode currents are doubled, while all DC electrode voltages remain the same as for single tube operation. If a cathode resistor is used, its value should be about one-half that for a single tube.

---

[2] One common, if not elegant, method of setting the balance of a push-pull audio stage is to adjust for "minimum hum."

Output Transformer

The load impedance is usually coupled to the tube of a class A power amplifier by means of a transformer [4]. This arrangement avoids passing the DC plate current through the load impedance and also makes it possible, by use of the proper turns ratio, to make any load present the desired impedance to the tube. The use of a transformer in this manner, however, causes the output of the amplifier to fall off at low and high frequencies. The extent of the decrease at the low end is determined largely by the inductance of the transformer primary; at the high end, the decrease is determined largely by the leakage inductance (in the case of a triode) or by the combination of leakage inductance and capacitance shunting the primary (in the case of a pentode or beam power tube).

The output transformer always delivers less energy to the load than is developed by the tube as a result of power losses in the device. Most of the transformer loss is caused by the resistance of the windings; flux densities are usually sufficiently low that core losses are negligible. The maximum current that an output transformer can carry is determined by the heating of the windings, while the maximum voltage that can be applied is limited by the permissible flux density in the core and the winding insulation rating. These factors operating together determine the power rating of the device.

Output transformers are physically much larger than power supply (60 Hz or 50 Hz) transformers of corresponding ratings, particularly output transformers that must respond to relatively low frequencies.

Many styles of audio output transformers have been developed and produced commercially. Wide variations in performance exist.

Class AB Power Amplifier

A class AB power amplifier employs two tubes connected in push-pull configuration with a higher negative grid bias than is used in a class A stage [2]. With this higher negative bias, the plate and screen grid voltages can usually be made higher than for class A amplifiers because the increased negative bias holds plate current within the limit of the tube plate dissipation rating. As a result of these higher voltages, more power output can be obtained from class AB operation.

Class AB amplifiers are subdivided into class AB_1 and class AB_2 as follows:

- In class AB_1, there is no flow of grid current. That is, the peak signal voltage applied to each grid is not greater than the negative grid bias voltage. The grids, therefore, are not driven to a positive potential and do not draw current.
- In class AB_2, the peak signal voltage is greater than the bias so that the grids are driven positive and draw grid current. Because of the flow of grid current in a class AB_2 stage, there is a loss of power in the grid circuit. Due to the large fluctuations of plate current in a class AB_2 stage, the plate power supply should have good regulation. Otherwise, fluctuations in plate current will cause fluctuations in the voltage output of the power supply, with the result that power output is decreased and distortion is increased.

Class AB operation is characterized by an efficiency performance intermediate between that of the corresponding class A and class B systems [4]. Thus, the plate efficiency that is realized in practice is commonly on the order of 35 to 50 percent. The DC plate current also is less when no signal is applied than when full signal is present, although the no-signal value is not nearly as small as for class B operation. Class AB amplifiers have been used to generate small and moderate levels of audio frequency power. They have, to some extent, displaced the true class A push-pull amplifier.

Class B Power Amplifier

A class B amplifier employs two tubes connected in push-pull configuration and biased so that plate current is almost zero when no signal voltage is applied to the grids [2]. Because of this low value of no-signal plate current, class B amplification has the same advantage of class AB_2—that is, high power output can be obtained without excessive plate dissipation. Class B operation differs from class AB_2 in that plate current is cut off for a larger portion of the negative grid swing, and the signal swing is usually larger than in class AB_2 operation.

Because certain triodes used as class B amplifiers are designed to operate very close to zero bias, the grid of each tube is at a positive potential during most or all of the positive half-cycle of its voltage swing. In this type of triode operation, considerable grid current is drawn and there is a loss of power in the grid circuit. This condition imposes similar requirements on the driver stage as in class AB_2 operation.

Power amplifier tubes designed for class A operation can be used in class AB_2 and class B service under suitable operating conditions. Several tube types are designed specifically for class B service. A characteristic common to all of these types is a high amplification factor. With a high amplification factor, plate current is small even when the grid bias is zero. These tubes, therefore, can be operated in class B service at a bias of zero volts so that no bias supply is required.

The frequency response characteristics of class B amplifiers depend upon the output transformer in much the same way as for class A amplifiers [4]. Thus, the falling off in amplification at low frequencies is determined by the primary inductance of the output transformer, while the leakage inductance controls the falling off in gain at high frequencies. The exact relationships existing in class B operation are, however, more complicated than in class A amplifiers because of the intermittent character of the plate current of the individual tubes. This fact also makes it desirable that the leakage inductance between the two halves of the primary winding be as small as possible.

Compared with class A power amplifiers, the class B arrangement has the advantage of higher plate efficiency, negligible power loss when no signal voltage is applied, and greater output power available from a given tube and plate supply system.

Cathode-Drive Circuits

The circuits described so far use vacuum tubes in a conventional grid-driven type of amplifier—that is, where the cathode is common to both the input and output circuits [2]. Tubes may also be employed as amplifiers in circuit arrangements that utilize the grid

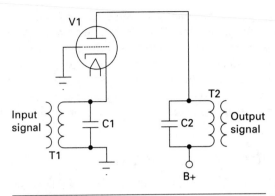

FIGURE 4.6   Cathode-drive circuit.
(*After* [2].)

or plate as the common terminal. Probably the most important of these amplifiers are the cathode-drive circuit and the cathode-follower circuit.

A typical cathode-drive circuit is shown in Figure 4.6. The load is placed in the plate circuit and the output voltage is taken between the plate and ground as in the grid-drive method of operation. The grid is grounded, and the input voltage is applied across an appropriate impedance in the cathode circuit. The cathode-drive circuit is particularly useful for high frequency operation in which it is necessary to obtain the low-noise performance usually associated with a triode but where a convention grid-drive circuit would be unstable because of feedback through the grid-to-plate capacitance of the tube. In the cathode-drive circuit, the grounded grid serves as a capacitive shield between the plate and cathode.

The input impedance of a cathode-drive circuit is approximately equal to $1/g_m$ when the load resistance is small compared to the plate resistance of the tube. A certain amount of power is required, therefore, to drive such a circuit.

An example of a cathode-follower circuit is shown in Figure 4.7. In this application, the load has been moved from the plate circuit to the cathode circuit of the tube.

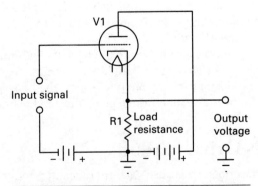

FIGURE 4.7   Cathode-follower circuit.
(*After* [2].)

The input voltage is applied between the grid and ground, and the output voltage is obtained between the cathode and ground. The voltage amplification of the circuit is always less than unity, and may be expressed by the following equations.
For a triode:

$$V.A. = \frac{\mu \times R_l}{r_p + [R_l \times (\mu + a)]}$$

For a pentode:

$$V.A. = \frac{g_m \times R_l}{1 + (g_m \times R_l)}$$

where:

μ = the amplification factor

R_l = load resistance in ohms

r_p = plate resistance in ohms

g_m = transconductance in mhos

The cathode follower permits the design of circuits that have high input resistance and high output voltage. The output impedance is quite low and very low distortion can be obtained. Cathode-follower circuits may be used for power amplifiers or as impedance transformers designed either to match a transmission line or to produce a relatively high output voltage at a low impedance level.

In a power amplifier that is transformer-coupled to the load, the same output power can be obtained from the tube as would be obtained from a conventional amplifier. The output impedance is very low and provides excellent damping to the load, with the result that very low distortion can be obtained.

When a cathode-follower circuit is used as an impedance transformer, the load is usually a simple resistance in the cathode circuit of the tube. With a relatively low value of cathode resistor, the circuit may be designed to supply significant power and to match the impedance of the device to a transmission line. With a somewhat higher value of cathode resistor, the circuit may be used to decrease the output impedance sufficiently to permit the output of audio signals along a line in which appreciable capacitance is present.

Phase Inverter

A phase inverter is a circuit used to provide resistance coupling between the output of a single tube stage and the input of a push-pull stage [2]. A phase inverter is necessary here because the signal-voltage input to the grids of a push-pull stage must be 180° out of phase and approximately equal in amplitude with respect to each other. Thus, when the signal voltage input to a push-pull stage swings the grid of one tube in a positive direction, it must swing the grid of the other tube in a negative direction by

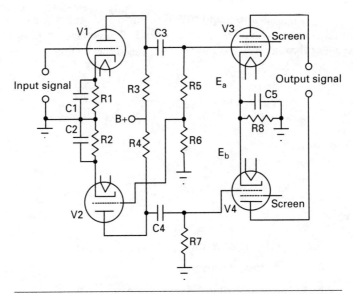

FIGURE 4.8   Use of a phase inverter to drive a push-pull power amplifier. (*After* [2].)

a similar amount. With transformer coupling between stages, the out-of-phase input voltage to the push-pull stage is supplied by means of a center-tapped secondary winding. With resistance coupling, the out-of-phase voltage is obtained by means of the inverter action of a tube. Figure 4.8 shows the use of a phase inverter to drive a push-pull power amplifier.

Inverse Feedback

An inverse-feedback circuit is one in which a portion of the output voltage of a tube is applied to the input of the same or a preceding tube in opposite phase to the signal applied to the tube [2]. There are two important advantages of such feedback:

- Reduced distortion for each stage included in the feedback circuit.
- Reduction in the variations in gain due to changes in power supply voltage, possible differences between tubes of the same type, or variations in the values of circuit constants included in the feedback circuit.

Inverse feedback is used in audio amplifiers to reduce distortion in the output stage where the load impedance on the tube is a loudspeaker. Because the impedance of a loudspeaker is not constant for all audio frequencies, the load impedance on the output tube varies with frequency. When the output tube is a pentode or beam power tube having high plate resistance, this variation in plate load impedance can, if not corrected, produce considerable distortion. Such distortion can be reduced by means of inverse feedback. Inverse-feedback circuits are of the constant-voltage type and the constant-current type.

Figure 4.9 illustrates the constant-voltage inverse feedback technique applied to a power output stage using a single beam power tube [2]. In this circuit, R2, R3, and C2 are connected as a voltage divider across the output of the tube. The secondary winding of the grid input transformer is returned to a point on this voltage divider. Capacitor C2 blocks the DC plate voltage from the grid. A portion of the tube AF output voltage is applied to the grid, approximately equal to the output voltage multiplied by

$$\frac{R_3}{R_2 + R_3}$$

This feedback voltage reduces the source impedance of the circuit and results in a decrease in distortion.

Referring to Figure 4.10, consider first an amplifier without inverse feedback. Assume that when a signal voltage e is applied to the grid, the AF plate current i_p has an irregularity in its positive half-cycle. This irregularity represents a departure from the waveform of the input signal and can therefore be considered distortion. For this plate current waveform, the AF plate voltage has a waveform shown by e_p. The plate voltage waveform is inverted compared to the plate current waveform because a plate current increase produces an increase in the drop across the plate load. The voltage at the plate is the difference between the drop across the load and the supply voltage. Thus, when plate current goes up, plate voltage goes down; when plate current goes down, plate voltage goes up.

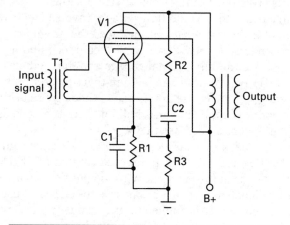

FIGURE 4.9   Power output stage using constant voltage inverse feedback. (*After* [2].)

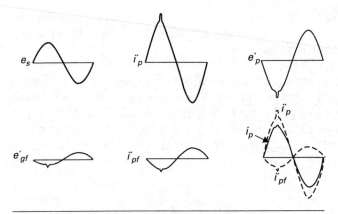

FIGURE 4.10   Voltage and current waveforms showing the effects of inverse feedback. (*After* [2].)

Next assume that inverse feedback is applied to the amplifier. The voltage fed back to the grid has the same waveform and phase as the plate voltage, but is smaller in magnitude. Hence, with a plate voltage of waveform shown by e'_p, the feedback voltage appearing on the grid is as shown by e'_{gf}. This voltage applied to the grid produces a component of plate current i''_{pf}. It is evident that the irregularity of the waveform of this component of plate current will act to cancel the original irregularity and thus reduce distortion.

After inverse feedback has been applied, the relations are as shown in the curve for i_p. The dotted curve shown by i''_{pf} is the component of plate current due to the feedback voltage on the grid. The dotted curve shown by i''_p is the component of plate current due to the signal voltage on the grid. The algebraic sum of these two components gives the resultant plate current shown by the solid curve of i_p. Because i''_p is the plate current that would flow without inverse feedback, it can be seen that the application of inverse feedback has reduced the irregularity in the output current. In this manner inverse feedback acts to correct any component of plate current that does not correspond to the input signal voltage and thus reduces distortion.

From the curves for i_p, it can be seen that, in addition to reducing distortion, inverse feedback reduces the amplitude of the output current. Consequently, when inverse feedback is applied to an amplifier there is a decrease in gain or power sensitivity as well as a decrease in distortion.

Inverse feedback may also be applied to resistance-coupled stages, as shown in Figure 4.11. The circuit is conventional except that feedback resistor R3 is connected between the plates of tubes V1 and V2. The output signal voltage of V1 and a portion of the output signal voltage of V2 appear across R3. Because the distortion generated in the plate circuit of V2 is applied to its grid out of phase with the input signal, the distortion in the output of V2 is comparatively low.

Inverse feedback circuits can also be applied to push-pull class A and class AB$_1$ amplifiers.

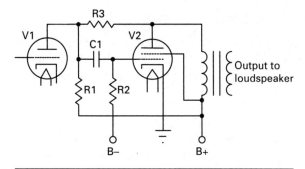

FIGURE 4.11   Resistance-coupled stage using a feedback resistor. (*After* [2].)

Constant-current inverse feedback is usually obtained by omitting the bypass capacitor across the cathode resistor. This method decreases gain and distortion, but increases the source impedance of the circuit. Consequently, the output voltage rises at the resonant frequency of the loudspeaker and accentuates *hangover* effects.

Inverse feedback is not generally applied to a triode power amplifier because the variation in speaker impedance with frequency does not produce much distortion in a triode stage having low plate resistance. It is sometimes applied in a pentode stage, but is not always convenient. As has been shown, when inverse feedback is used in an amplifier, the driving voltage must be increased in order to provide full power output. When inverse feedback is used with a pentode, the total driving voltage required for full power output may be inconveniently large, although still less than that required for a triode.

Because a beam power tube gives full power output on a comparatively small driving voltage, inverse feedback is especially applicable to beam power tubes. By means of inverse feedback, the high efficiency and high power output of beam power tubes can be combined with freedom from the effects of varying speaker impedance.

Corrective Filter

A corrective filter can be used to improve the frequency characteristics of an output stage using a beam power tube or a pentode when inverse feedback is not applicable [2]. The filter consists of a resistor and a capacitor connected in series across the primary of the output transformer. Connected in this way, the filter is in parallel with the plate load impedance reflected from the voice coil to the output transformer. The magnitude of this reflected impedance increases with increasing frequency in the middle and upper audio frequency range. The impedance of the filter, however, decreases with increasing frequency. It follows that by use of the proper values for the resistance and the capacitance in the filter, the effective load impedance on the output tubes can be made reasonably constant for all frequencies in the middle and upper range. The result is an improvement in the frequency characteristics of the output stage.

Tone Control

A tone control is a variable filter, or one in which at least one element is adjustable, that can be controlled by the user to change the frequency response of an amplifier. Three common approaches to tone control have evolved in the consumer marketplace:

- Single-control tone adjustment.
- Two-control tone adjustment—one to control low-frequency response (*bass*) and the other to control high-frequency response (*treble*).
- Graphic equalizer, where a number of controls are used to adjust specific frequency bands.

For the purposes of this book, we will focus on the two-control method.

A two-stage tone-control network is shown in Figure 4.12 [2]. Separate bass and treble controls are provided. Figure 4.13*a* shows simplified representations of the bass control when the potentiometer is turned to its extreme variations—that is, full *boost* and *cut*. In this network, the parallel resistor-capacitor (RC) combination is the controlling factor. For bass boost, capacitor C2 bypasses resistor R3 so that less impedance is placed across the output to the grid of V2 at high frequencies than at low frequencies. For bass cut, the parallel combination is shifted so that C1 bypasses R3, causing more high-frequency output than low-frequency output. Essentially, the network is a variable-frequency voltage divider. With proper selection of component values, the circuit may be made to respond to changes in the R3 potentiometer setting only for frequencies below about 1000 Hz.

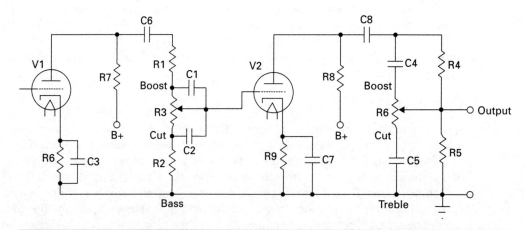

FIGURE 4.12 A two-stage tone-control stage incorporating separate bass and treble controls. (*After* [2].)

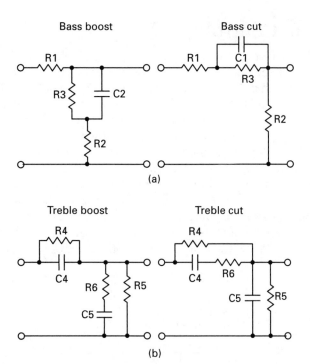

FIGURE 4.13   Simplified representation of the tone-control circuit illustrated in Figure 4.12 at extreme ends of the potentiometer range: (*a*) bass control circuit, (*b*) treble control circuit. (*After* [2].)

Figure 4.13*b* shows extreme positions of the bass and treble controls. The attenuation of the circuits is approximately the same at 1000 Hz. The treble boost circuit relies on C4 to bypass R4 for high-frequency signals, thereby increasing the high-end frequency response. In the treble cut circuit, the parallel RC elements serve to attenuate the high-frequency signal voltage as capacitor C5 bypasses the resistance across the output. The effect of the capacitor is negligible at low frequencies; above 1000 Hz, however, the signal voltage is attenuated at a maximum rate of 6 dB per octave.

In an amplifier, the tone control network is usually inserted after the first AF amplifier tube and before the power output stage(s). General design rules include the following:

- If the amplifier incorporates negative feedback, the tone control may be inserted in the feedback network; otherwise, it should be connected to a part of the amplifier that is external to the feedback loop.
- The overall gain of a well-designed tone-control network should be approximately unity.

Volume Control

It is always necessary to provide a means of controlling the gain of an audio amplifier so that the output can be kept at the desired level irrespective of the signal voltage applied at the input [4]. The standard method of controlling volume manually, shown in Figure 4.14*a*, makes use of a grid leak in the form of a high-resistance potentiometer. In such an arrangement, the changes in the frequency response characteristic that occur with volume settings are small and in the nature of a slight improvement in the high-frequency response at low-volume settings as a result of the reduced resistance across which the input capacitance of the output tube is connected. This arrangement also avoids direct current flow through the volume control potentiometer, a requirement that must be met to avoid noise when the control is adjusted.

It is also possible to control the volume of a transformer-coupled amplifier stage by means of a potentiometer connected across the transformer secondary, as shown in Figure 4.14*b*. Such an arrangement has the disadvantage, however, of causing the

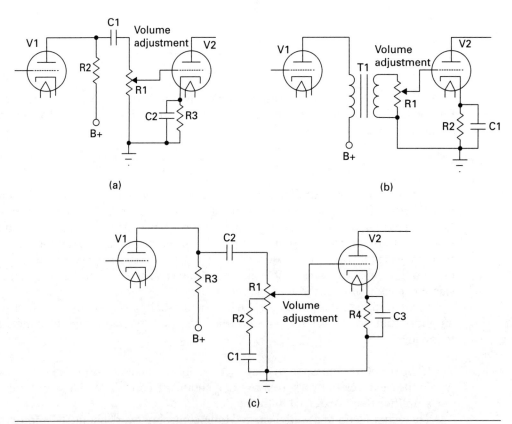

(a)

(b)

(c)

FIGURE 4.14 Common methods of controlling volume in an audio amplifier: (*a*) resistance-coupled amplifier, (*b*) transformer-coupled amplifier, (*c*) tone-compensated circuit. (*After* [4]. *Used with permission.*)

frequency response characteristic to vary considerably with the gain setting. As a result, it is customary to control the gain of an audio frequency amplifier in one of the resistance-coupled stages.

The characteristics of the ear are such that the apparent loudness of low-frequency tones relative to middle- and high-frequency tones is less as the volume level of reproduction is reduced. In order to correct for this effect, volume controls are sometimes arranged so that, as the volume control reduces the gain, the reduction is less for low frequencies than for middle- and high-frequency components. Such an arrangement is termed a tone-compensated volume control. A typical example is shown in Figure 4.14c.

The volume control in an amplifier must always be placed at a relative low power level point in the system. Otherwise, the first stage(s) of amplification could be overloaded when large input signals are present.

Noise in Electronic Devices

All electronic circuits are affected by any number of factors that cause their performance to be degraded from the ideal assumed in simple component models in ways that can be controlled but not eliminated entirely [5]. One limitation is the failure of the model to account properly for the real behavior of components, either the result of an oversimplified model description or variations in manufacture. Usually, by careful design, the engineer can work around the limitations of the model and produce a device or circuit whose operation is very close to predictions. One source of performance degradation that cannot be overcome easily, however, is noise.

When vacuum tubes first came into use in the early part of the twentieth century, they were extensively used in radios as signal amplifiers and for other related signal conditioning functions. Thus, the measure of performance that was of greatest importance to circuit designers was the quality of the sound produced at the radio speaker. It was immediately noticed that the sound coming from speakers not only consisted of the transmitted signal but also of popping, crackling, and hissing sounds, which seemed to have no pattern and were distinctly different from the sounds that resulted from other interfering signal sources, such as other radio stations using neighboring frequencies. This patternless or random signal was labeled "noise," and has become the standard term for signals that are random and are combined with the circuit signal to affect the overall performance of the system.

As the study of noise has progressed, engineers have come to realize that there are many sources of noise in circuits. The following definitions are commonly used in discussions of circuit noise:

- **White noise** Noise whose energy is evenly distributed over the entire frequency spectrum, within the frequency range of interest. Because noise is totally random, it may seem inappropriate to refer to its frequency range, because it is not really periodic in the ordinary sense. Nevertheless, by examining an oscilloscope trace of white noise, one can verify that every trace is different—as the noise never repeats itself—and yet each trace looks the same. Similarly, an analog TV set tuned to an unused frequency displays never-ending "snow" that always looks the same, yet clearly is always changing. There is a strong theoretical foundation to represent the frequency content of such signals as covering the frequency spectrum evenly. In this way the impact on other periodic signals can be analyzed. The term "white noise" arises from the fact that—similar to white light, which has equal amounts of all light frequencies—white noise has equal amounts of noise at all frequencies within circuit operating range.

- **Interference**   The name given to any predictable, periodic signal that occurs in an electronic circuit in addition to the signal the circuit is designed to process. This component is distinguished from a noise signal by the fact that it occupies a relatively small frequency range, and because it is predictable, it can often be filtered out. Interference usually results from another electronic system, such as an interfering radio source.

- **Thermal noise**   Any temperature-dependent noise generated within a circuit. This signal usually is the result of the influence of temperature directly on the operating characteristics of circuit elements, which—because of the random motion of molecules as a result of temperature—in turn creates a random fluctuation of the signal being processed.

- **Shot noise**   A type of circuit noise that is not temperature-dependent, and is not white noise in the sense that it tends to diminish at higher frequencies. (See Figure 4.15.) This noise usually occurs in components whose operation depends on a mean particle residence time for the active electrons within the device. The cutoff frequency above which noise disappears is closely related to the inverse of this characteristic particle residence time. It is called "shot noise" because in

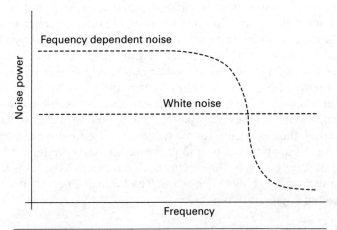

FIGURE 4.15   A plot of signal power versus frequency for white noise and frequency-limited noise. (*After* [3].)

a radio it can make a sound similar to buckshot hitting a drum, as opposed to white noise, which tends to sound more like hissing (because of the higher frequency content).

Thermal Noise

By definition, thermal noise is internally generated noise that is temperature-dependent [5]. While first observed in vacuum tube devices (because their amplifying capabilities tend to bring out thermal noise), it is also observed in semiconductor devices. It is a phenomenon resulting from the ohmic resistance of devices that dissipate the energy lost in them as heat. Heat consists of random motion of molecules, which are more energetic as temperature increases. Because the motion is random, it is to be expected that as electrons pass through the ohmic device incurring resistance, there should be some random deviations in the rate of energy loss. This fluctuation has the effect of causing variation in the resulting current, which is noise. As the temperature of the device increases, the random motion of the molecules increases, and so does the corresponding noise level.

The noise level produced by thermal noise sources is not necessarily large; however, because source signal power may also be low, it is usually necessary to amplify the source signal. Because noise is combined with the source signal and both are then amplified, with more noise added at each successive stage of amplification, noise can become a noticeable phenomenon.

Hum

Circuit noise related to the AC line operating frequency (60 Hz or 50 Hz) is usually grouped into a category of noise simply referred to as *hum*. While there can be a number of causes for hum in a vacuum tube amplifier; the most common ones are

- Poor power supply regulation
- Heater circuits
- Lack of proper grounding of sensitive circuits
- Magnetic radiation from power supply inductors

Some of these causes are more easily addressed than others.

In the case of heater connections, the leads to the tube sockets are typically twisted tightly and kept away from high-impedance circuits. Other techniques include using a center-tapped filament transformer where the center tap is grounded or connected to a positive voltage source.

Magnetic fields induce voltages in coils and in loops that may exist if the wiring is improperly arranged. The principal sources of magnetic fields in the vicinity of an amplifier are the power transformer and the leads carrying the filament or heater currents to the tubes [4]. Power transformers should be placed as far as possible from portions of the system sensitive to magnetic fields, particularly input transformers and associated tubes. Stray magnetic fields can be minimized by using a power transformer with low leakage flux. It is also helpful to employ a nonmagnetic chassis.

Input transformers present a particularly difficult problem in high-gain audio amplifiers because they operate at the point in the system where the power level is the lowest. The input transformer of a high-gain audio amplifier must be magnetically shielded and placed as far as possible from the power transformer, with an orientation that experiment indicates will minimize hum pickup.

Electrostatic fields can cause trouble with parts of the amplifier having a high impedance to ground, since any electrostatically induced current flowing to ground will produce a hum voltage that is proportional to the impedance between the part and the circuit ground. To minimize electrostatic pickup, low-level stage tubes should be shielded and all leads kept as short as possible.

Microphonics

The arrangement of elements within a vacuum tube is such that, when exposed to vibration, one or more of the elements may move in relation to the other elements. Such effects are usually minor and can be largely ignored, except for high gain stages working with very low signal levels. Different tube types are more or less sensitive to this problem. Vibrations can be coupled to the tube via movement of the chassis or through the vibration of components mounted on the chassis, such as the power transformer. In the transformer case, hum can be generated in a circuit through mechanical vibration of an audio input stage tube.

Solutions to microphonic issues usually involve mechanical isolation of the tube socket from the chassis, typically through use of a socket designed to reduce mechanical coupling (e.g., rubber pads or other shock-absorbing material). Individual tubes of the same type may vary considerably in their tendency to microphonic action.

Shielding

In an audio amplifier, shielding between the input and output circuits must be carefully considered; the higher the gain of an amplifier, the more important the shielding. Particular care should be taken with the lead dress of input and output circuits so as to minimize the possibility of stray coupling. Shielding of low-level audio conductors within the chassis is recommended.

High-Voltage Power Supplies

Virtually all vacuum tubes require one or more sources of high voltage to operate. The usual source of operating power is a single-phase supply operating from AC line current. The primary AC-to-DC converting device may be a rectifier tube or silicon rectifier(s).

Silicon Rectifier

Rectifier parameters generally are expressed in terms of reverse-voltage ratings and mean forward-current ratings in a ½-wave rectifier circuit operating from a 60 Hz supply and feeding a purely resistive load [1]. The three principal reverse-voltage ratings are

- **Peak transient reverse voltage** (V_{rm}) The maximum value of any nonrecurrent surge voltage. This value must never be exceeded.
- **Maximum repetitive reverse voltage** $(V_{RM(rep)})$ The maximum value of reverse voltage that may be applied recurrently (in every cycle of 60 Hz power). This includes oscillatory voltages that may appear on the sinusoidal supply.
- **Working peak reverse voltage** $(V_{RM(wkg)})$ The crest value of the sinusoidal voltage of the AC supply at its maximum limit. Rectifier manufacturers generally recommend a value that has a significant safety margin, relative to the peak transient reverse voltage (V_{rm}), to allow for transient overvoltages on the supply lines.

There are three forward-current ratings of similar importance in the application of silicon rectifiers:

- **Nonrecurrent surge current** $(I_{FM.surge})$ The maximum device transient current that must not be exceeded at any time. $I_{FM.surge}$ is sometimes given as a single value, but more often is presented in the form of a graph of permissible surge-current values versus time. Because silicon diodes have a relatively small thermal mass, the potential for short-term current overloads must be given careful consideration.
- **Repetitive peak forward current** $(I_{FM,rep})$ The maximum value of forward current reached in each cycle of the 60 Hz waveform. This value does not include random peaks caused by transient disturbances.
- **Average forward current** $(I_{FM,av})$ The upper limit for average load current through the device. This limit is always well below the repetitive peak forward-current rating to ensure an adequate margin of safety.

Rectifier manufacturers generally supply curves of the instantaneous forward voltage versus instantaneous forward current at one or more specific operating temperatures. These curves establish the forward-mode upper operating parameters of the device.

Figure 4.16 shows a typical rectifier application in a bridge circuit.

Rectifier Tube

The two most common vacuum tube rectifier configurations, half-wave and full-wave, are illustrated in Figure 4.17. In the half-wave circuit, current flows through the tube to the filter on every other half-cycle of the AC input voltage when the plate is positive

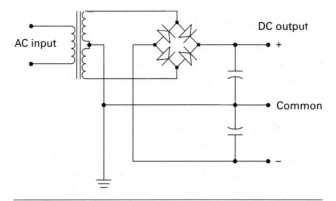

FIGURE 4.16 Conventional capacitor input filter full-wave bridge.

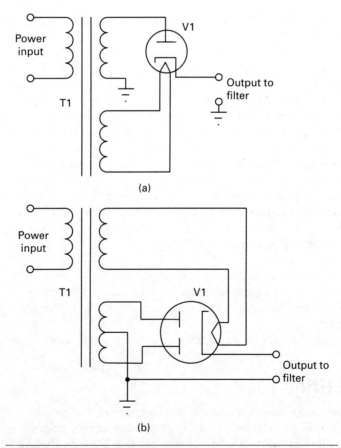

FIGURE 4.17 Vacuum tube rectifier circuits: (*a*) half-wave rectifier, (*b*) full-wave rectifier. (*After* [2].)

with respect to the cathode [2]. In the full-wave circuit, current flows to the filter on every half-cycle through plate number 1 on the half-cycle when plate number 1 is positive with respect to the cathode and through plate number 2 on the next half-cycle when plate number 2 is positive with respect to the cathode. Because the current flow to the filter is more uniform in the full-wave circuit than in the half-wave circuit, the output of the full-wave circuit requires less filtering.

Parallel operation of rectifier tubes provides an output current greater than that obtainable with the use of one tube. The permissible voltage and load conditions per tube are the same, but the total load-current-handling capability of the rectifier is approximately doubled.

Two rectifier tubes can also be connected in parallel to yield somewhat increased output voltage by virtue of the lower combined internal resistance of the paired devices. Stabilizing resistors may be necessary, depending on the tube type and the circuit. Losses in the resistors will impact any gain in output voltage.

Figure 4.18 illustrates the process of converting an AC voltage into a DC voltage.

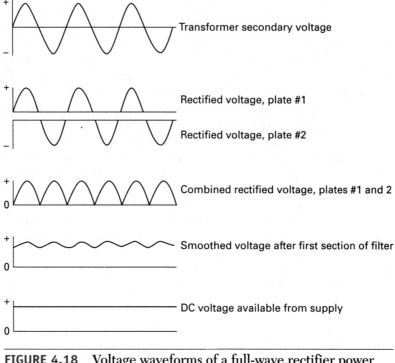

FIGURE 4.18 Voltage waveforms of a full-wave rectifier power supply. (*After* [2].)

Power Supply Filter Circuits

The filter network for a high-voltage power supply typically consists of a series inductance or resistance and one or more shunt capacitances. Bleeder resistors and various circuit protection devices are usually incorporated as well. Filter systems can be divided into two basic types:

- **Inductive input** Filter circuits that present a series inductance to the rectifier output
- **Capacitive input** Filter circuits that present a shunt capacitance to the rectifier output

Figure 4.19 illustrates some of the various possible configurations.

Inductive Input Filter

The inductive input filter is the most common configuration found in high-voltage equipment [1]. When the input inductance is infinite, current through the inductance is constant and is carried at any moment by the rectifier anode that has the most positive voltage applied to it at that instant. As the alternating voltage being rectified passes through zero, the current suddenly transfers from one anode to another, producing square current waves through the individual rectifier devices.

When the input inductance is finite (but not too small), the current through the input inductance tends to increase when the output voltage of the rectifier exceeds the average or DC value, and to decrease when the rectifier output voltage is less than the DC value. If the input inductance is too small, the current decreases to zero during a portion of the time between the peaks of the rectifier output voltage. The conditions then correspond to a capacitor input filter system.

The output waveform of the rectifier can be considered as consisting of a DC component upon which are superimposed AC voltages (ripple voltages). To a first approximation, the fluctuation in output current resulting from a finite input inductance can be considered as the current resulting from the lowest-frequency component of the ripple voltage acting against the impedance of the input inductance. This assumption is permissible because the higher-frequency components in the ripple voltage are smaller, and at the same time encounter higher impedance. Furthermore, in practical filters the shunting capacitor following the input inductance has a small impedance at the ripple frequency compared with the reactance of the input inductance. The peak

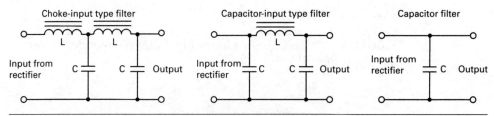

FIGURE 4.19 Typical power supply filters. (*After* [2].)

current resulting from a finite input inductance, therefore, is given approximately by the relation [4]

$$\frac{I_f}{I_i} = 1 + \frac{E_1}{E_0}\frac{R_{eff}}{\omega L_1}$$

where:

I_f = peak current with finite input inductance

I_i = peak current with infinite input inductance

E_1/E_0 = ratio of lowest-frequency ripple component to the DC voltage in the rectifier output

R_{eff} = effective load resistance

ωL_1 = reactance of the incremental value of the input inductance at the *lowest* ripple frequency

This equation is derived as follows:

- The peak alternating current through the input inductance is approximately $E_1/\omega L_1$
- The average or DC current is E_0/R_{eff}
- The peak current with finite inductance is, therefore: $(E_1/\omega L_1) + (E_0/R_{eff})$
- The current with infinite inductance is E_0/R_{eff}

The effective load resistance value consists of the actual load resistance plus filter resistance plus equivalent diode and transformer resistances.

The normal operation of an inductive input filter requires that there be a continuous flow of current through the input inductance. The peak alternating current flowing through the input inductance, therefore, must be less than the DC output current of the rectifier. This condition is realized by satisfying the approximate relation

$$\omega L_1 = R_{eff}\frac{E_1}{E_0}$$

In the practical case of a 60 Hz single-phase full-wave rectifier circuit, the foregoing equation becomes

$$L_1 = \frac{R_{eff}}{1130}$$

The minimum allowable input inductance (ωL_1) is termed the *critical inductance.* When the inductance is less than the critical value, the filter acts as a capacitor input circuit. When the DC drawn from the rectifiers varies, it is still necessary to satisfy the ωL_1 equation at all times, particularly if good voltage regulation is to be maintained. To fulfill this requirement at small load currents without excessive inductance, it is necessary to place a bleeder resistance across the output of the filter system in order to limit R_{eff} to a value corresponding to a reasonable value of L_1.

Capacitive Input Filter

When a shunt capacitance rather than a series inductance is presented to the output of a rectifier, the behavior of the circuit is greatly modified [1]. Each time the positive crest alternating voltage of the transformer is applied to one of the rectifier anodes, the input capacitor charges up to just slightly less than this peak voltage. The rectifier then ceases to deliver current to the filter until another anode approaches its peak positive potential, when the capacitor is charged again. During the interval when the voltage across the input capacitor is greater than the potential of any of the anodes, the voltage across the input capacitor drops off nearly linearly with time because the first filter inductance draws a substantially constant current from the input capacitor. A typical set of voltage and current waves is illustrated in Figure 4.20.

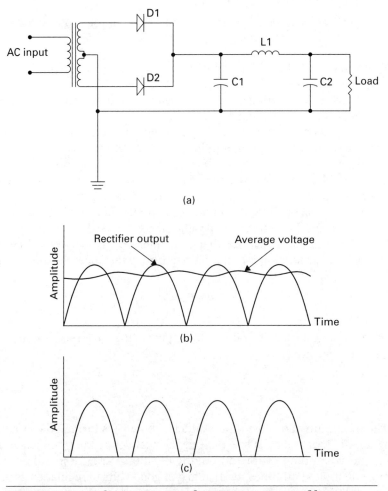

FIGURE 4.20   Characteristics of a capacitive input filter circuit: (*a*) schematic diagram, (*b*) voltage waveshape across the input capacitor, (*c*) waveshape of current flowing through the diodes. (*From* [1]. *Used with permission.*)

The addition of a shunt capacitor to the input of a filter produces fundamental changes in behavior, including the following:

- The output voltage is appreciably higher than with an inductance input.
- The ripple voltage is less with a capacitance input filter than with an inductance input filter.
- The DC voltage across the filter input drops as the load current increases for the capacitive input case instead of being substantially constant, as for the inductive input case.
- The ratio of peak-to-average anode current at the rectifiers is higher in the capacitive case.
- The utilization factor of the transformer is less with a capacitive input configuration.

Filters incorporating shunt capacitor inputs generally are employed when the amount of DC power required is relatively small. Inductance input filters are used when the amount of power involved is large; the higher utilization factor and lower peak current result in important savings in rectifier and transformer costs under these conditions.

Grid Voltage

Grid voltage is applied to make the grid negative with respect to the cathode by a specified amount [2]. Grid voltage may be obtained from one of three primary sources:

- From a fixed power supply source, known as *fixed bias*
- From the voltage drop across a resistor in the cathode circuit, known as *cathode bias* or *self-bias*
- From the voltage drop across a resistor in the grid circuit, known as *grid-resistor bias*

The cathode-biasing method utilizes the voltage drop produced by the cathode current flowing through a resistor connected between the cathode and the negative terminal of the power supply. The cathode current is equal to the plate current in the case of a triode, or to the sum of the plate and grid #2 currents in the case of a tetrode, pentode, or beam power tube. Because the voltage drop along the resistance is increasingly negative with respect to the cathode, the required negative grid-bias voltage can be obtained by connecting the grid return to the negative end of the resistance.

Bypassing of the cathode-bias resistor depends on the circuit design requirements. In AF circuits, the use of an unbypassed resistor will reduce distortion by introducing degeneration into the circuit; however, it will also decrease gain and power sensitivity. When bypassing is used, it is important that the bypass capacitor be sufficiently large so as to have negligible reactance at the lowest frequency to be amplified.

In the case of a power-output tube having high transconductance, such as a beam power tube, it may be necessary to shunt the bias resistor with a small-value capacitor in order to prevent oscillations.

The use of a cathode resistor to obtain bias voltage is not recommended for amplifiers in which there is appreciable shift of electrode currents with the application of a signal. For such amplifiers, a separate fixed supply is recommended.

The grid-biasing method is also a self-bias approach because it utilizes the voltage drop across the grid resistor produced by a small amount of grid current flowing in the grid-cathode circuit. This current is due to

- The electromotive potential difference between the materials comprising the grid and cathode
- Grid rectification when the grid is driven positive

A large value of resistance is required to limit this current to a very small value and to avoid undesirable loading effects on the preceding stage.

Screen Voltage Supply

The positive voltage for the screen grid (grid #2) may be obtained from a tap on a voltage divider from the plate supply, a series resistor connected to the high-voltage source, or a separate fixed supply [2]. The screen grid voltage for a tetrode should not be obtained from a series resistor from the high-voltage source because of the characteristic screen-grid variations of tetrodes.

When pentode or beam power tubes are operated under conditions where a large shift of plate and screen grid currents does not take place with the application of the signal, the screen grid voltage may be obtained through a series resistor from the high-voltage source (Figure 4.21). This method of supply is possible because of the high uniformity of the screen grid current characteristics in pentodes and beam power tubes.

It is important to note that the plate voltage of tetrodes, pentodes, and beam power tubes should be applied before or simultaneously with the screen grid voltage. Otherwise, with voltage on the screen grid only, the screen grid current may rise high enough to cause excessive screen grid dissipation.

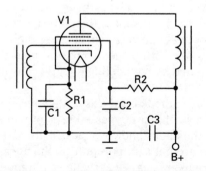

FIGURE 4.21   Pentode circuit in which screen grid voltage is supplied through a series resistor. (*After* [2].)

Supply Decoupling

Filters are typically included in the voltage supply leads of each tube in order to return the signal through a low impedance path directly to the cathode rather than by way of the voltage supply circuit [2]. Figure 4.22 illustrates several types of filter circuits. Capacitor C forms the low impedance path, while the choke or resistor assists in diverting the signal through the capacitor by offering a high impedance to the power supply circuit.

The choice between a resistor and a choke depends mainly upon the permissible DC voltage drop through the filter. In circuits where the current is small, resistors are practical; where the current is large or regulation is important, chokes are more suitable.

Protection Measures

Vacuum tubes are designed to withstand considerable abuse [1]. The maximum ratings for most devices are conservative. Typically, the maximum dissipation for each grid indicated on the data sheet should not be exceeded except for time intervals of less than 1 s or so. The maximum dissipation rating for each grid structure is usually considerably above typical values used for maximum output so that ample operating reserve is provided. The time of duration of overload on a grid structure is necessarily short because of the small heat-storage capacity of the grid wires. Furthermore, grid temperatures cannot be measured or seen, so no warning of accidental overload is apparent.

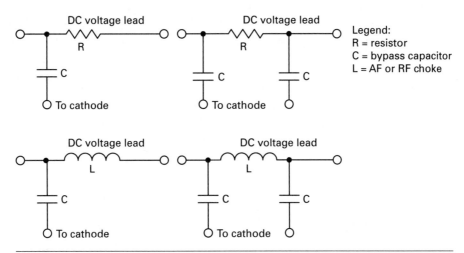

FIGURE 4.22   Typical power supply decoupling (filter) circuits. (*After* [2].)

Chapter 5

Interconnection, Layout, and Operating Environment

The design of a reliable, high-performance audio system requires careful consideration of how the system interfaces with its surroundings—both electrical and environmental. The choices made during the design phase will impact—positively or negatively—the performance and reliability of the overall system. The methods used to interconnect various pieces of equipment, and the hardware used to make those interconnections, determine in part how the overall system will operate. Likewise, the construction method chosen will impact the complexity and cost of the project.

Wiring Practices

All signal transmission media impair—to some extent—an input electrical signal as it moves from one point to another [1]. Foremost among these impairments is attenuation. The distance over which transmission is possible is determined technically by the threshold sensitivity of the signal receiver. Subjectively, the maximum distance is determined by user-established specifications for tolerable signal bandwidth reduction and signal to noise ratio (SNR) increase. Noise in this analysis is a generic term that includes Gaussian noise present in all active components in the transmission system, unwanted signals (crosstalk) coupled from parallel signal transmission circuits, electromagnetic interference (EMI), and radio frequency interference (RFI) from the total environment through which the signal passes. All other transmission impairments can be grouped within the generic term of nonlinearities. These include passband frequency response flatness deviations, harmonic distortion, and aberrations detected as frequency-specific differences in signal gain and phase.

Crosstalk between cables carrying different types and levels of signals can be minimized by isolating the cables into separate groups for video, audio, control, data, and power. Audio cable should be further subdivided into the following categories:

- Low level (below –20 dBm)
- Medium level (–20 to +20 dBm)
- High level (above +20 dBm)

Types of Noise

Open (noncoaxial) wiring can couple energy from external fields [1]. These fields result from power lines, signal processes, and radio frequency sources. The extent of coupling is determined by the following:

- Loop area between conductors
- Cable length
- Cable proximity
- Frequency
- Field strength

Two basic types of noise can appear on AC power, audio, video, and computer data lines within a device or facility—*normal mode* and *common mode*. Each type has a particular effect on sensitive load equipment. The normal-mode voltage is the potential difference that exists between pairs of power or signal conductors. This voltage also is referred to as the *transverse-mode* voltage. The common-mode voltage is a potential difference (usually noise) that appears between power or signal conductors and the local ground reference.

The common-mode noise voltage will change depending upon what is used as the ground reference point. It is often possible to select a ground reference that has a minimum common-mode voltage with respect to the circuit of interest, particularly if the reference point and the equipment are connected by a short conductor. Common-mode noise can be caused by electrostatic or electromagnetic induction.

Interfaces

Analog audio devices of all types have used the "RCA" connector for line-level inputs and outputs for many decades. We will follow that practice in the projects contained later in this book. For amplifier outputs, a number of different approaches have been taken over the years. For the projects described in this book, we will use barrier strips.

Balanced and Unbalanced Systems

While rarely used in consumer audio gear, nearly all professional audio systems use balanced inputs and outputs because of the noise immunity they afford [1]. These techniques, although not utilized in the projects contained in this book, are nonetheless instructive insofar as noise control is concerned.

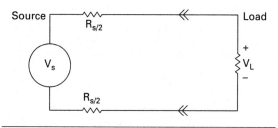

FIGURE 5.1 A basic source and load connection. No grounds are indicated, and both the source and the load float. (*After* [1].)

Figure 5.1 shows a basic source and load connection. No grounds are present, and both the source and load float. This is the optimum condition for equipment interconnection.

Either the source or the load may be tied to ground with no problems, provided only one ground connection exists. Unbalanced systems are created when each piece of equipment has one of its connections tied to ground, as shown in Figure 5.2. This condition occurs if the source and load equipment have unbalanced (single-ended) inputs and outputs. This type of equipment uses chassis ground for one of the audio conductors. This is a common practice in consumer audio products. Equipment with unbalanced inputs and outputs is less expensive to build, but when various pieces of audio gear are tied together, noise problems can result. These problems are compounded when the equipment is separated by a significant distance.

As shown in Figure 5.2, a difference in ground potential causes current flow in the ground wire. This current develops a voltage across the wire resistance, and the ground noise voltage adds directly to the signal itself. Because the ground current usually results from leakage in power transformers and line filters, the 60-Hz signals give rise to hum. Reducing the wire resistance through a heavier ground wire helps reduce the hum, but cannot eliminate it completely.

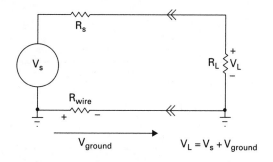

FIGURE 5.2 An unbalanced system in which each piece of equipment has one of its connections tied to ground. (*After* [1].)

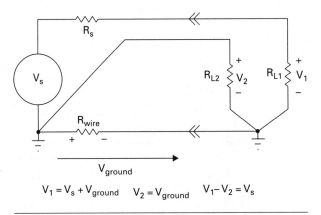

$$V_1 = V_s + V_{ground} \qquad V_2 = V_{ground} \qquad V_1 - V_2 = V_s$$

FIGURE 5.3   Illustration of how ground loop noise can be canceled by amplifying both the high side and the ground side of the source and subtracting the two signals. (*After* [1].)

By amplifying both the high side and the ground side of the source and subtracting the two to obtain a difference signal, it is possible to cancel the ground loop noise (see Figure 5.3). This is the basis of the differential input circuit. Unfortunately, problems still can exist with the unbalanced source to balanced load system shown in the drawing. The reason centers around the impedance of the unbalanced source. One side of the line will have a slightly lower amplitude because of impedance differences in the output lines.

By creating an output signal that is out of phase with the original, a balanced source can be created to eliminate this error (see Figure 5.4). As an added benefit, for

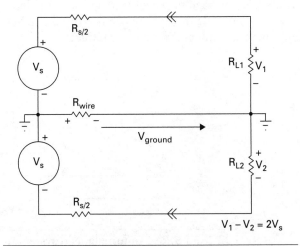

$$V_1 - V_2 = 2V_s$$

FIGURE 5.4   A balanced source configuration where the inherent amplitude error of the system shown in Figure 5.3 is eliminated. (*After* [1].)

a given maximum output voltage from the source, the signal voltage is doubled over the unbalanced case.

The added complexity of the balanced input and output circuits must be weighed against the possibility of noise introduced because of the electrical environment of the interconnection lines, and the length of those lines. Balanced circuits can also be achieved through the use of line-level audio transformers. Transformers offer a simple solution, although they are usually more expensive than the integrated circuit (IC) solutions readily available for active balanced systems. The performance limitations that may be imposed by transformers is another consideration.

Physical Layout Considerations

The mechanical layout of a vacuum tube amplifier is dictated by its basic operating parameters. The largest physical items must be accommodated first. For a power amplifier, this is typically the power transformer and output transformer(s). Common practice involves locating these devices as far as physically possible from low-level input stages. In addition to distance, metal shielding may be used. After the transformers have been placed, the vacuum tubes can be located, usually in a logical progression from low to high signal levels. Power output tubes are invariably placed adjacent to their associated output transformer. A typical amplifier chassis layout is shown in Figure 5.5.

An amplifier with two or more stages should be constructed with a straight-line layout so that maximum separation is provided between the signal input and output

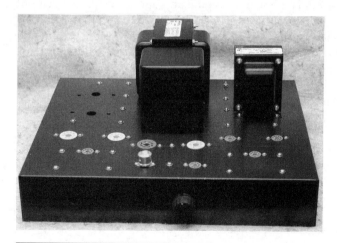

FIGURE 5.5   Audio amplifier chassis under construction. Note the placement of the transformers and tube sockets.

circuits and terminals. Power-carrying connections, particularly line AC voltages, should be isolated as far as possible from signal connections, especially from the input connections. Signal-carrying conductors, even when shielded, should not be cabled together with power supply conductors. Internal chassis wiring of AC-operated tube heaters, switches, pilot light sockets, and other devices should be twisted and placed flat against the chassis. All connections to ground should be made securely to the ground plane on the circuit board or to the chassis itself.

Because power amplifiers and high-voltage power supplies normally dissipate large amounts of heat, they should be constructed and installed to ensure adequate ventilation for the tubes and other components. For units that are installed horizontally (with the tubes vertical), ventilation holes may be provided to allow cooling air to enter the chassis from below and up through the chassis to cool the tubes and other components.

Hand-Wired or PWB?

The choice of a printed wiring board (PWB) or hand-wired terminal strips is a major consideration in the construction of an audio device. The PWB offers a cleaner overall appearance, although it also requires more up-front engineering. The impact of heat from the tubes on a PWB must be considered. Over time, heat can damage a PWB. A variety of techniques are available to avoid this type of damage, which is more critical on power devices (which invariably generate more heat).

Hand-wired terminal strips are simple and effective, and require a minimal amount of advanced planning. In some cases, both PWB and hand-wiring may be used.

It should be noted that some audiophiles claim they can hear the difference between an amplifier built using a PWB and one that is hand-wired with terminal strips. Technical justification for such a claim is difficult to imagine. The author will take no position on this question, other than to point out that he cannot hear any difference. If you can, that's all that counts.

Table 5.1 compares some of the more obvious merits of each approach. While not an exhaustive examination, the table nonetheless provides a starting point for further consideration.

There is a viable intermediate point between a hand-wired chassis and a PWB, which utilizes an insulating board of perhaps 2-in high by 5-in wide with "tags" installed at the outer edges of the long dimension. With these "tag boards" (sometimes referred to as "turret boards"), the passive components are connected as needed using the opposing rows of tags. Component leads or wires connect the tag board circuits to the tube sockets. Tag boards are available in a variety of sizes and may be customized for a particular application.

TABLE 5.1   Relative Merits of Common Wiring Methods

Parameter	Terminal Strip	Printed Wiring Board	Hybrid
Up-front engineering effort	Minimal: For a simple amplifier, a hand-drawn sketch will usually suffice.	Considerable: The circuit diagram must be transferred to a PWB layout mask with the dimensions of the components taken into consideration. It should be noted that software is readily available for this task.	Considerable: About the same as for a PWB.
Cost	Minimal: The cost for the terminal strips is a small percentage of the overall bill of materials.	Considerable: The cost for the PWB can be a significant percentage of the overall bill of materials. Builders can make their own boards using readily available supplies and chemicals, or utilize one of a number of prototyping houses that specialize in short-run jobs.	Considerable: About the same as for a PWB.
Reliability	Very good: Once properly constructed the circuits will operate for a very long time with no maintenance (parts failures notwithstanding).	Good: If not subjected to excessive heating from tubes, resistors, and other components, the reliability of a PWB is nearly as good as a hand-wired unit. If, however, the tube sockets are mounted directly on the PWB, the life expectancy of the PWB may be limited to some extent. Likewise, if heat-generating devices such as high-wattage resistors are not given sufficient clearance above the PWB, the lifetime of the board may be limited.	Very good: This assessment assumes that the objective of the hybrid approach is to place the tubes on the metal chassis and use short interconnecting wires to the PWB. This further assumes that heat-producing components are separated from the PWB by some means. These assumptions are the basis for using a hybrid design.

(continued)

TABLE 5.1 Relative Merits of Common Wiring Methods (*Continued*)

Parameter	Terminal Strip	Printed Wiring Board	Hybrid
Maintainability	Good: Replacing a failed component can usually be accomplished with minimal impact to other components. Devices that connect at a terminal with three or more component leads present can, however, be problematic. In this case, in order to successfully replace one component, it may be necessary to replace other components at the same time because of the likelihood of damage or limited lead length.	Good: Replacing a failed component can usually be accomplished with very little impact to other devices on the board. This assessment, however, assumes the use of extreme care when working with the board, including using only enough soldering iron heat to accomplish the task. One common problem with replacing devices on a PWB involves traces or trace holes lifting up from the board, which requires careful repair.	Very good: The considerations outlined for hand-wired and PWB devices apply here as well.
Appearance	Classic: If the intent is a classic tube look, hand-wired is clearly the way to go.	Production: This approach usually gives the look of a production-run consumer product. Perhaps not the best look for a classic tube design. Then again, if the objective is a high-tech look, this is the way to go.	Potentially classic: If done correctly, the hybrid approach can give the best of both the hand-wired and PWB techniques. This assessment assumes that the tubes are mounted on the metal chassis and the PWB is mounted underneath the chassis, using connecting wires from the tube sockets to the PWB(s).

(*continued*)

TABLE 5.1   Relative Merits of Common Wiring Methods (*Continued*)

Parameter	Terminal Strip	Printed Wiring Board	Hybrid
Construction time requirements	Medium: With a layout strategy mapped out, the time required to install parts is not excessive.	Low: With proper documentation, a PWB can be populated with components in a short period.	High: Since the hybrid method involves a combination of both hand-wired and PWB approaches, the time required to construct the device is usually greater than using either single approach.
Skill level required	Medium: While a novice can produce a well-executed device, this method is really intended for the experienced hobbyist.	Low: Assuming good soldering skills and proper documentation, minimal construction experience is typically required to populate a PWB.	High: Mating the chassis-mounted devices to the PWB(s) adds a new level of complexity to the project.
Reproducibility	Low: Building the second amplifier will take about as long as building the first.	High: If the intent is to build multiple amplifiers, the up-front engineering efforts can be recouped in subsequent projects.	Medium: As a combination of hand-wired and PWB approaches, the gains for producing multiple units are typically minimal.
Overall performance	Very good: The hand-wired technique typically results in short lead lengths and use of the chassis itself as a ground plane. As such, performance issues are rarely associated with this method.	Good: The primary limitation of a PWB in power amplifier applications is the integrity of the ground plane. Layout software typically does a good job of providing for sufficient ground-potential real estate; however, more than one iteration of the board may be necessary to achieve the desired result.	Good: In addition to the considerations noted for PWB designs, the issue of the interconnecting wires from the tube sockets to the PWB must be taken into account. At audio frequencies, the chances of spurious products and/or circuit instabilities are usually small.

Chassis

Although it usually imparts no sonic qualities to the end product, the selection of the chassis and front panel can—and often does—have a considerable impact on the cost of the project. The basic options are

- Off-the-shelf chassis and front panel, with the necessary holes and openings punched as needed
- Custom chassis designed to fit a particular project

It is no surprise that the latter option is almost always the more expensive. To keep costs to something reasonable, the projects contained in this book will utilize off-the-shelf materials, with custom components used only when necessary.

The front panel can represent a considerable investment, depending on the sophistication of the desired finished product. The labeling of controls is important to the appearance and utility of the device. Making the panel look professionally done may require using professional services.

One fundamental decision that needs to be made early on in the design process is the desired "look" of the unit. The three most common options and their related considerations are listed in Table 5.2.

The considerations listed in Table 5.2 are not exhaustive, but should provide a starting point for further consideration.

Thermal Properties

Adequate cooling of the tube envelope and seals is one of the principal factors affecting vacuum tube life [2]. Deteriorating effects increase directly with the temperature of the tube envelope and seals. Inadequate cooling is almost certain to invite premature failure of the device. Tubes operated at VHF and above are inherently subjected to greater heating action than tubes operated at lower frequencies. This results directly from the following:

- The flow of larger RF charging currents into the tube capacitances at higher frequencies
- Increased dielectric losses at higher frequencies
- The tendency of electrons to bombard parts of the tube structure other than the normal elements as the operating frequency is increased

Greater cooling, therefore, is required at higher frequencies. The technique most commonly used to cool such devices is forced air cooling. For audio applications, where the ambient room noise is a major consideration, the use of fans is unusual and strongly discouraged. Fortunately, the operating considerations for tubes at high frequencies do not apply for most audio applications.

TABLE 5.2   Considerations Regarding the Chassis/Cabinet "Look and Feel"

Consideration	*Classic* Look	*Modern* Look	*Don't Care* Look
Definition	An appearance that matches the styling of the 1950s and '60s. Attributes could include 1) Tubes visible to the user 2) Black or chrome chassis 3) Knob styles of the time period 4) Analog meters (if used) 5) Incandescent status lights	What the user thinks constitutes a "modern" look. Wide latitude exists here, since "modern," like beauty, is in the eye of the beholder. Attributes could include 1) Tubes visible to the user 2) Black chassis and front panel 3) Unique knobs 4) Digital metering (if used) 5) LED status lighting	Since the unit will not be seen, the way it looks doesn't matter
Cost impact	Medium, although a chrome-plated chassis will drive up the cost	High, particularly if digital metering is desired	Low
Engineering detail	Low impact	High impact, particularly if digital metering and features such as remote control are desired	Low impact
Overall appearance	Very nice	Very nice	Who cares? Nobody will ever see it.

It is important to note that some tubes, notably rectifiers and beam power tubes, should be mounted vertically. It is possible to mount some of these devices in a horizontal position if certain guidelines are followed; for example, pins 1 and 5 in the vertical plane. Generally speaking, however, it is best to mount tubes vertically and provide sufficient separation between devices to allow for adequate cooling.

Heat Transfer Process

In the commonly used model for materials, heat is a form of energy associated with the position and motion of the molecules, atoms, and ions of the material [3]. The position is analogous with the state of the material and is potential energy, while the motion of the molecules, atoms, and ions is kinetic energy. Heat added to a material makes it hotter, and heat withdrawn from a material makes it cooler. Heat energy is measured in calories (cal), British thermal units (Btu), or joules. One calorie is the amount of energy required to raise the temperature of one gram of water one degree Celsius (14.5 to 15.5°C). One Btu is the unit of energy necessary to raise the temperature of one pound of water by one degree Fahrenheit. One joule is an equivalent amount of energy equal to the work done when a force of one newton acts through a distance of one meter.

Temperature is a measure of the average kinetic energy of a substance. It can also be considered a relative measure of the difference of the heat content between bodies.

Heat transfers through a material by conduction resulting when the energy of atomic and molecular vibrations is passed to atoms and molecules with lower energy. As heat is added to a substance, the kinetic energy of the lattice atoms and molecules increases. This, in turn, causes an expansion of the material that is proportional to the temperature change over normal temperature ranges. If a material is restrained from expanding or contracting during heating and cooling, internal stress is established in the material.

Thermal energy may be transferred by any of three basic modes:

- Conduction
- Convection
- Radiation

Conduction

Heat transfer by conduction in solid materials occurs whenever a hotter region with more rapidly vibrating molecules transfers a portion of its energy to a cooler region with less rapidly vibrating molecules [3]. Conductive heat transfer is the most common form of thermal exchange in electronic equipment. Thermal conductivity for solid materials used in electronic equipment spans a wide range of values, from excellent (high conductivity) to poor (low conductivity). Generally speaking, metals are the best conductors of heat, whereas insulators are the poorest.

With regard to vacuum tube devices, conduction is just one of the cooling mechanisms involved in maintaining element temperatures within specified limits. At high power levels, a secondary mechanism works to cool the device. For example, heat generated within the tube elements is carried to the cooling surfaces of the device by conduction. Convection or radiation then is used to remove heat from the tube itself.

Convection

Heat transfer by natural convection occurs as a result of a change in the density of a fluid (including air), which causes fluid motion [3]. Convective heat transfer between a heated surface and the surrounding fluid is accompanied by a mixing of fluid adjacent to the surface. Vacuum tube devices relying on convective cooling invariably utilize air passing by (or in the case of high-power transmitting tubes, through) the hot envelope.

Radiation

Cooling by radiation is a function of the transfer of energy by electromagnetic wave propagation [3]. The wavelengths between 0.1 and 100 m are referred to as thermal radiation wavelengths. The ability of a body to radiate thermal energy at any particular wavelength is a function of the body temperature and the characteristics of the surface of the radiating material.

Reliability Considerations

Long-term reliability of a vacuum tube requires regular attention to the operating parameters and environment. The first step in this process is the design and layout of the unit.

Failure Mechanisms

Components and systems never fail without a reason. The reason may be difficult to find, but the determination of failure modes and weak areas in a given system is fundamental to increasing the reliability—whether for a large and complex system or for a small and simple one.

All equipment failures are logical; some are predictable. A system failure usually is related to poor-quality components or to abuse of the system or a part within, either because of underrating or environmental stress [4]. Even the best-designed components can be badly manufactured. A process can go awry, or a step involving operator intervention may result in an occasional device that is substandard or likely to fail under normal stress. Hence, the process of screening and/or *burn-in* to weed out problem parts is a universally accepted quality control tool for achieving high reliability.

Figure 5.6 illustrates what is commonly known as the *bathtub curve*. It divides the expected lifetime of a class of parts into three segments: infant mortality, useful life, and wear-out. A typical burn-in procedure for a passive device or semiconductor device consists of the following steps:

- The parts are electrically biased and loaded; that is, they are connected in a circuit representing a typical application.
- The parts are cycled on and off (power applied, then removed) for a predetermined number of times. The number of cycles may range from ten to several thousand during the burn-in period, depending on the component under test.
- The components under load are exposed to high operating temperatures for a selected time (typically 72 to 168 hours). This constitutes an accelerated life test for the part.

An alternative approach for passive devices and semiconductor devices involves temperature shock testing, in which the component product is subjected to temperature extremes, with rapid changes between the hot-soak and cold-soak conditions. After the stress period, the components are tested for adherence to specifications. Parts meeting

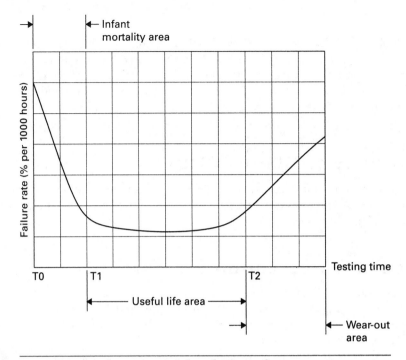

FIGURE 5.6 The statistical distribution of equipment or component failures versus time. (*After* [4].)

the established specifications are accepted for shipment to customers. Parts that fail to meet them are discarded.

For a vacuum tube, production testing is a much simpler process of running the device at the rated operating values for some specified period and confirming adherence to specifications. Stress testing is usually counterproductive.

A systems approach to reliability is effective, but not foolproof. The burn-in period is a function of statistical analysis; there are no absolute guarantees. The natural enemies of electronic parts are heat, vibration, and excessive voltage. Not all of these parameters are under the control of the designer—and ultimately end user— all of the time.

Failure Modes

Normal wear issues aside, the circuit elements most vulnerable to failure in any piece of electronic hardware are those exposed to the outside world [4]. In most systems, the greatest threat typically involves one or more of the following components or subsystems:

- The AC-to-DC power supply
- Sensitive signal-input circuitry
- High-power output stages and devices

Derating of individual components is a key factor in improving the overall reliability of a given system. The goal of derating is the reduction of electrical, mechanical, thermal, and other environmental stresses on a component to decrease the degradation rate and prolong expected life. Through derating, the margin of safety between the operating stress level and the permissible stress level for a given part is increased. This adjustment provides added protection from system overstress that is unforeseen during design.

The average life that may be expected from a vacuum tube is a function of many operational parameters, including

- Filament voltage
- Ambient operating temperature
- Output power
- Operating efficiency

The best estimate of life expectancy for a given system comes from on-site experience.

Vacuum Tube Life

A vacuum tube suffers wear-out because of a predictable chemical reaction [4]. The cathode is the heart of any tube. The device is said to wear out when filament emissions are inadequate for rated output or acceptable waveform distortion. Among the factors that impact the lifetime of a vacuum tube are the quality of the tube vacuum and the operating temperature of the filament. In high-power vacuum tubes, the filament voltage is typically regulated to maximize lifetime. This practice is limited to high-power (and high-priced) devices, although the idea of trying it in an audio application is interesting.

The design of electron tubes allows for some variation in the voltage and current supplied to the filament or heater, but most satisfactory results are obtained from operating at the rated values [5]. When the voltage is too low, the temperature of the cathode is below normal, with the result that electron emission may be limited. The limited emission may cause unsatisfactory operation and reduced tube life. On the other hand, high heater voltage may cause rapid evaporation of cathode material and shortened tube life.

To ensure proper tube operation, measure the filament or heater voltage at the socket terminals while the equipment is in the normal operating mode. Care must be taken to prevent excessive voltage drop in the chassis wiring that would result in low voltage at the tube terminals.

Apart from wear-out failure, catastrophic failures can occur and are usually divided into two primary categories:

- The short-circuiting of broken or warped elements within the tube
- A loss of vacuum in the device

Air in the tube causes a loss of dielectric standoff between the internal elements. The end results of an interelectrode short-circuit or loss of vacuum are about the same—catastrophic failure of the device, and perhaps one or more passive components in the related circuit.

Fault Protection

Catastrophic failures of tubes and other devices (such as electrolytic capacitors) occur from time to time. Whenever possible, designs should anticipate the potential for failure and include mechanisms to eliminate (or at least reduce) the possibility of collateral damage. Such mechanisms range from simple (installing properly sized fuses) to complex, such as current-sensing circuits that remove power when a preset trip point has been exceeded.

Thermal Cycling

Thermal cycling of a vacuum tube is inevitable. Each cycle causes the elements to expand and then (when turned off) contract. The materials used in tubes are designed to withstand these forces. The only practical advice on limiting the potential of thermal cycling damage includes (please forgive the obvious nature of this advice):

- Do not needlessly turn the system on and off.
- Minimize the potential for excessive heating of the tube and related hardware through management of the ambient operating temperature to the extent possible.

Shipping and Storing Vacuum Tubes

Because of their fragile nature, vacuum tubes should packed for shipment in foam-filled or other shock-absorbing containers [4]. When it is necessary to transport a tube from one location to another, use the original packing material if available. Tubes should be removed whenever the system is shipped. Never leave a tube installed in its socket during an equipment move.

Never allow a tube to roll along a surface. Damage to the filament or to other elements in the device may result.

Chapter 6

Construction Project Considerations

Any electronics project begins with a plan regarding what to build and how to build it. For simple projects, the level of planning can be minimal. Larger projects, however, particularly ones in which expensive items are required, deserve careful planning. A key element of planning involves developing the necessary documentation that will guide construction and ultimately facilitate maintenance.

Planning the Project

Project planning is a science in itself. We will not explore the many facets of this discipline, but instead focus on key steps necessary for building electronics devices. The process of developing a project plan yields the following critical information:

- The feature set of the finished product—what the product is intended to do and the designed performance level
- A bill of materials (BOM) for the project and the approximate cost for the hardware
- A list of work that must be completed by one or more outside vendors, such as sheet metal work, and the necessary drawings for those vendors
- A detailed schematic of the system and separate drawings showing the placement of major components
- An estimate of the time needed to complete the project

In the process of compiling this information, it is likely that potential problems will be identified and addressed before any holes are punched in sheet metal or any components are purchased. In the long run, careful planning shortens the development time and reduces the final cost of the project by identifying unforeseen issues early on.

Finding the Necessary Parts

Once the BOM has been developed, the task of finding the needed parts can begin. For projects such as a vacuum tube amplifier, it is unlikely that all necessary parts can be acquired from a single vendor. While it is certainly most efficient to order from the smallest number of vendors possible, it may be necessary to order specialty items, such as transformers and tubes, from particular sources. Common components, such as resistors, capacitors, hardware, and semiconductors, can usually be ordered from one of the major catalog parts houses.[1] Specialty items may require some research to locate.[2] Deal with companies that offer such services as fast delivery and online information regarding in- and out-of-stock items. Buy from companies that offer easy return privileges for unacceptable or defective parts.

While ordering the parts for a project may seem a simple and relatively quick step, keep in mind that some parts may not be readily available. It is likely that some parts will be out of stock and therefore require a significant lead-time for delivery. For out-of-stock items, it may be appropriate to select a different manufacturer in order to complete the order. For example, a given on-line retailer typically offers the same (or similar) components from different manufacturers. It may be possible, therefore, to get the part needed without having to resort to a backorder situation.

Of all the parts needed to build an audio amplifier, the transformers will likely be the most expensive. The vacuum tubes can also be expensive, depending on the type chosen. Audio output transformer options tend to divide into the following groups:

- "Universal replacement" types that are intended for general-purpose applications where high audio fidelity is not necessarily required. These types provide good value for noncritical applications.
- High-quality types designed specifically for exceptional frequency response. These types are more expensive, but are recommended for critical applications. Construction techniques are employed that optimize performance, notably extended frequency response.
- Potted ultra-linear types designed for exceptional performance and appearance. These are predictably the most expensive audio transformers available, but also provide the highest performance.

Prices for vacuum tubes vary widely from one make to the next. While it is not always clear why one device might cost three times more than the same type from a different manufacturer, the following general classifications tend to group tubes into price categories:

- **New old stock (NOS)**   These tubes, as the name implies, are devices manufactured many years (or even decades) ago. Because receiving tubes have a long shelf life when properly handled, these devices are a good choice.

---

[1] In the United States two such suppliers are Allied Electronics (www.alliedelec.com) and Newark Electronics (www.newark.com).

[2] Online vendors that specialize in vacuum tubes include The Tube Store (http://thetubestore.com/), Tube Depot (http://tubedepot.com/), Tube World (www.tubeworld.com/), VacuumTubes.Net (www.vacuumtubes.net/), and Antique Electronic Supply (www.tubesandmore.com/).

- **New**   Devices recently manufactured by one of a number of tube companies around the world. In some cases, the tubes have been optimized for one particular application or another.
- **Match pair**   Intended for push-pull audio circuits, match pair devices are used for demanding applications. In any product run, small variations in materials and construction will result in minor differences in operating characteristics. With a matched pair, the two devices are essentially identical, which results in better performance.

The choices available for transformers and tubes, thus, serve a wide range of preferences, needs, and budgets.

The effort required to locate the necessary parts for a project should not be underestimated. The two main challenges are: 1) the part specified in the original tube design is no longer available and so a substitute must be identified, and 2) sorting out the part numbers and vendors can consume a large amount of time. In writing this book, the author likely spent more time finding and ordering parts than actually installing them. For this reason, each project in this book includes a detailed parts list that includes manufacturer part numbers and stock numbers from one of the major parts houses. It is hoped this information will substantially shorten the amount of time readers spend on acquiring the right components.

In developing parts lists for the projects that follow, the author has—in a number of cases—observed considerable variation in cost for the same basic part. For example, a 22 μF 450 V capacitor from one manufacturer might cost $3.50 retail from one vendor, with the same value device available from another vendor for $35.00. It is unclear why one device is ten times more costly than the other. It may be worth the price; that is a decision each reader can make on his or her own. For the projects described in this book, the author gravitated toward the more reasonably priced alternative.

Sheet Metal Work

For any project builder who does not own sheet metal–working equipment (that would probably be most readers of this book), getting the necessary chassis and front panel work completed can be a daunting task. A search of the Web will turn up specialty fabrication houses that will do *one-off* work, but that can get very expensive very fast. The best solution is probably to find a sheet metal shop in your local area that is willing to take on an occasional one-off project.[3]

Designing an equipment chassis from the ground up can yield exceptional results. The costs can also be exceptional. In many cases, however, it makes sense to begin with stock sheet metal components, such as a chassis and front panel, and have them modified (punched/drilled) to accommodate the necessary parts. When dealing with a local sheet metal fabricator, it is often helpful to mark the holes that need to be made

---

[3] Two fabrication houses you may want to consider are Metal Dynamix of Mount Laurel, NJ, (www.connectworld.net/rrdata/aluminum-fabrication.html) and Fabcon of Santa Ana, CA (www.fabcon.com/index.html).

on an adhesive overlay (or with a marker pen) and ask the operator to "make it like this." For a complex layout, detailed drawings will probably be required.

Sheet metal components, of course, can be machined to requirements using only a drill press and set of hole punches. A drill press is recommended over a hand drill because of the accuracy with which the holes can be made. Always use a set punch to keep the drill bit from "walking" over the metal. Hole punches, or "chassis punches," come in a variety of sizes, typically from 0.5-inch to 2-inches or more in diameter. For most electronics projects using vacuum tubes, several sizes will be needed, including the following:

- 0.5-inch punch for switches and variable resistors, and for cable entry
- 0.75-inch punch for nine-pin miniature tubes (exact sizes vary)
- 1-inch punch for eight-pin tubes (exact sizes vary, as do mounting options)

Chassis punches are available from a number of sources[4] in a wide variety of sizes. The most economical approach may be to purchase a set of five or more punches of common sizes. When purchasing such tools, note that sets are available specified by the size of the hole cut and by the size of the electrical conduit through which a hole cut by the punch can pass. For example, a hole punch for a 0.5-inch conduit will make a 0.75-inch hole. Look instead for the actual hole size.

If a clean finished product is desired, it is essential to plan ahead for chassis work. Start by purchasing samples of the devices and hardware needed for the project. This will identify any special considerations for particular components, and identify tools (punches, drill bits, etc.) needed to complete the project. Figure 6.1 shows various chassis preparation steps.

The front panel of the unit can present the greatest challenge because it is what most people will see most of the time. The front panel must provide mechanical support for user-operated controls and include legends for those controls (e.g., "volume," "power," etc.). While press-apply legends are easy to use and inexpensive, the finished look may be less than desired. As with the sheet metal work, a local vendor may be able to help.

Although more expensive than a do-it-yourself front panel, vendors are available that specialize in short-run or prototype front panels. The clear benefits here are a professional appearance and durable finish. For projects in this book, the author used the following services, which provided good results:[5]

- **Screen-printed polycarbonate overlay** This vendor (Metalphoto of Cincinnati, www.mpofcinci.com/) will work with the customer to design a custom panel or work from a customer-supplied drawing file (AutoCAD, Adobe Illustrator, and other formats). The overlay is printed and then affixed to a metal backer (the chassis front panel), and the necessary openings (holes) are punched in the overlay.

  A wide variety of options are available, including turnkey services. The polycarbonate

---

[4] Greenlee Tools is a major provider of such tools; see www.greenlee.com.
[5] There are other vendors in this space as well.

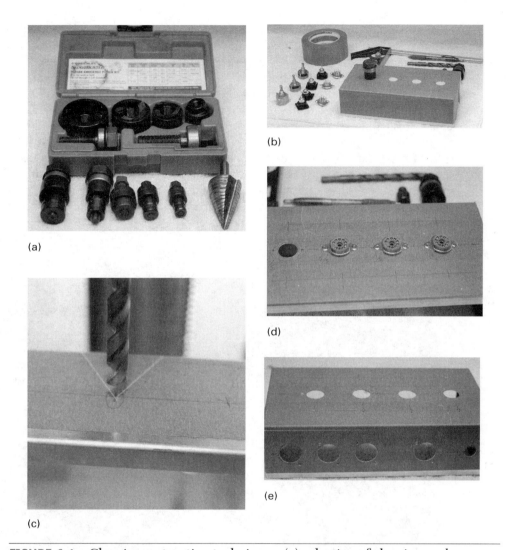

FIGURE 6.1 Chassis construction techniques: (*a*) selection of chassis punches for use with tube-based projects; (*b*) the chassis is covered with masking tape to facilitate marking of hole positions and sizes for the components; (*c*) use a drill press for precision cuts and control over the workpiece; (*d*) after the holes are punched, check for proper fit; (*e*) completed chassis.

overlay is quite durable and can include any number of legends and colors. This approach provides a professional appearance for all types of instruments.

- **Engraved panel** This vendor (Front Panel Express, www.frontpanelexpress .com/index.html) offers a free software package that allows the user to design chassis panels (or an entire enclosure) and transmit the design to the company, which then provides a bid and ultimately the finished product. Options include

1) the type, finish, and thickness of the panel; 2) legend fill colors; 3) holes and openings; and 4) custom engravings. This approach provides a professional appearance for consumer electronics products.

For a do-it-yourself approach, the chassis and front panel will, invariably, need to be painted. Make sure and use the right paint for the job—one that will adhere to the metal finish and provide good scratch-resistance. Here again, a local vendor may be the best way to go.

Many painting options are available. Spray paint is readily available in a number of colors and textures. Use of a primer is advisable. The chassis and other metal components can typically be purchased finished or unfinished. Try various paints on sample pieces of metal before settling on a final selection. Some painting considerations are illustrated in Figure 6.2.

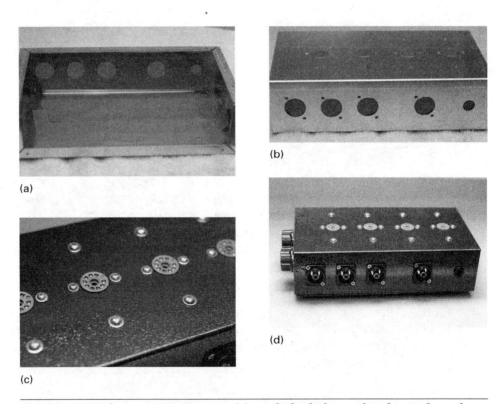

(a)

(b)

(c)

(d)

FIGURE 6.2 Chassis painting steps: (*a*) mask the holes in the chassis from the component side to prevent overspray, (*b*) make sure the surface is clean and dry prior to painting, (*c*) place components on the chassis, (*d*) finished chassis

Printed Wiring Boards

If your project will include one or more PWBs, the fabrication options are 1) home-brew using off-the-shelf chemicals and supplies, or 2) a vendor specializing in prototype boards. A web search will turn up several suppliers in this space, some offering free component layout software to develop the board.[6] With this approach, the vendor takes the user-designed layout and runs any number of boards. A variety of high-end production options are available, including silk-screening of component layout and other data. Such services, of course, increase the price but do provide an excellent end result.

When designing a project, it can be cost-effective to develop one or more basic layouts that can be duplicated as needed. For example, a stereo amplifier typically includes two identical signal paths. It may be less expensive to have two identical boards made than one larger, more complex board.

Placing Components

The first step in designing a PWB layout is to consider the signal flow and physical constraints, such as heat-producing items and components that may be unusually heavy and require additional support [1]. Some components, because of their size and/or weight, are more efficiently placed on the chassis and connected using hook-up wire. It is usually best practice to place parts only on one side of the board. Some component locations will be dictated by items such as connectors, switches, and other devices that mount to the chassis or front panel. Keep the trace lengths to a minimum. Group parts by circuit stage to facilitate short interconnecting traces. Arrange devices in an organized manner, usually based on an X-Y grid.

Some circuits require a number of wire connections from devices such as transformers. While it may be logical to group all of the connections in one area of the board, it may not be very efficient in terms of PWB traces. As a practical matter, straight connections between distant parts of a board are much more difficult to handle than connections involving components, which by their nature provide break points in the trace and allow other traces to cross.

Connecting wires from devices such as transformers and other components not mounted on the PWB to the board can be made by soldering the leads to pads on the board. This has the benefit of simplicity, but makes maintenance difficult since it may not be possible to view the underside of the PWB without unsoldering the leads. As an alternative, board-mounted barrier strips may be used, or so called "quick-disconnect" terminals. For boards with a large number of connections, the "quick-disconnect" method is advisable.

Position polarized parts (e.g., diodes and electrolytic capacitors) with the positive leads all having the same orientation, if possible. It may also be helpful to use a square pad to mark the positive leads of these components. Allow generous space between components to make the trace routing easier.

Most layout software will include a component library of common parts, which probably will not include vacuum tubes. Specialty components can be made by

---

[6] One such vendor is ExpressPCB, www.expresspcb.com.

TABLE 6.1 Trace Width as a Function of Current Carrying Capability (*After* [1].)

Trace Width (inches)	Current Carrying Capability
0.010	0.3 Amps
0.015	0.4 Amps
0.020	0.7 Amps
0.025	1.0 Amps
0.050	2.0 Amps
0.100	4.0 Amps
0.150	6.0 Amps

placing a series of individual pads and then grouping them together. Correct pin spacing is, of course, important; a caliper or other measuring device may be used.

After placing all the components, it may be worthwhile to print out a copy of the layout and position the devices to confirm there is sufficient space for all parts. After the components have been placed, the next step is to lay the power, ground, and signal traces. Do not snake or daisy-chain power rails from one section or part to the next. A substantial ground plane will minimize conducted and radiated noise, particularly on high-impedance, low-level circuits.

Various strategies can be used with regard to trace layout. Structures known as *vias* (a plated-through hole) are commonly used to move signals from one layer to another. Traces that carry significant current should be wider than signal traces. Table 6.1 provides rough guidelines on how wide to make a trace for a given amount of current. Remember that the maximum possible current through any given trace will be determined by the thinnest section of the trace. As noted, for lines carrying appreciable current (above 2 A), the traces become wide and, therefore, are difficult to handle efficiently on the board. For high-current circuits, it may be best to use jumper wires instead.

Consider the space between traces and adjacent traces or pads. Small spaces can lead to a short circuit when soldering components in place. Additional space must be provided for high voltages. Text can be placed on the top ("silkscreen") layer to show component layouts and connection points if desired. The information contained on the silkscreen layer is helpful when populating the board.

The primary application envisioned for PWB design software is solid-state–based devices. Vacuum tube circuits are probably a bit of a deviation from the usual run of boards. As such, it is important to size the board elements to accommodate the voltage and current levels associated with tubes. The following general guidelines are recommended as a starting point:

- Default clearance around traces = 0.05 inches, minimum
- Default clearance around pads = 0.05 inches, minimum
- Default via = 0.115 inch round with 0.052-inch hole
- Minimum trace width = 0.025 inch (for currents of 1 A or less)

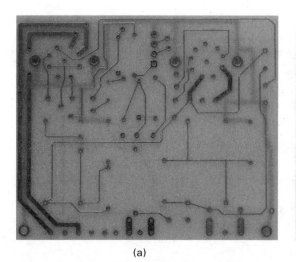 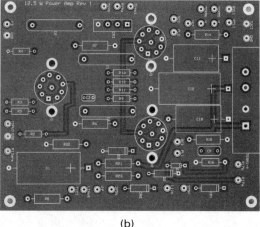

(a)   (b)

FIGURE 6.3   PWB manufacturing options: (*a*) basic board with the necessary traces, vias, and holes; (*b*) a premium board with the addition of a silkscreen overlay and solder mask.

Manufacturing options for a PWB include 1) a basic board with the necessary traces, vias, and holes; and 2) a premium board with the addition of a silkscreen overlay showing component layout and a solder mask on one or both sides of the board. Not surprisingly, option #2 is more expensive—perhaps a 2× premium. Either approach will, of course, work. It may be worthwhile to use option #1 for initial development and option #2 once the board design has been finalized. This assumes, of course, that you plan to build more than one unit, which may or may not be the case.[7] Figure 6.3 shows example boards based on option #1 and option #2.

While PWB vendors can assist in addressing specific needs, it is recommended that designers use the available tools and processes rather than request special, one-time changes. For example, in designing the PWB for the preamplifier described in Chapter 9, the author found that mounting a particular type of potentiometer would require a special step in the manufacturing process. In addition to increasing the cost of the board, making the change would increase the production lead time. The author, instead, chose a different component—one in which mounting could be accommodated without the need to drill a special-size hole in the board.

Tools

The simplest construction or maintenance task is next to impossible without the right tools [2]. There is more to a well-stocked tool chest than meets the eye. While tools may seem too basic to even warrant discussion, having the right tools is essential to

---

[7] The author can supply PWBs for certain circuits contained in this book. See details in Chapter 13.

TABLE 6.2 Basic Tools and Supplies for
Electronics Construction and Maintenance

Bench lamp

Compressed air canister

Crimping tool

Desoldering braid

Fine- and medium-grade solder

Flux remover spray

Magnifying glass

Nutdriver set

Pliers and wire cutters

PWB holding fixtures

Screwdriver set

Solder flux

Soldering irons (35 W and 45 W)

Spray-type contact cleaner

Tweezers

Wire stripper

efficient equipment construction and/or repair. Table 6.2 lists common tools, fixtures, and chemicals that should be on hand in any shop.

Deciding whether to invest in a customized tool kit or to buy the tools individually is not always an easy decision. Although at first glance it might seem more economical to buy tools one at a time as needed, a careful examination of the needs could point to a different conclusion. Manufacturers offer packages intended for particular applications that include tools which may or may not be readily available from your local parts supplier.

High-quality standard tool kits for hobbyists and professionals are available from a number of reputable suppliers. Because of the competitive nature of the tool kit market, buyers usually receive good value on any of these standard packages.

Case style and construction are important points to consider when buying a tool kit. It makes little sense to economize by buying an inexpensive case if it does not provide adequate protection for the contents. The case is the backbone of the tool kit. Style and construction play a major role in assuring its long-term functionality. When selecting a case, consider whether it will hold items other than tools, such as test instruments, electronic components, and documentation.

Consider also whether the case will be used in a single location or transported. A tool case that will be riding in the back of a pickup truck must be more substantial than one that is seldom moved. If the case is properly selected, tools and test equipment will be organized for easy access. A good tool pallet does more than

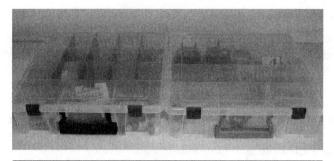

FIGURE 6.4   Storage containers for organizing project components

simply help keep track of tools; it provides the most frequently used tools in the most accessible locations. Delicate tools should be protected, and the weight of the tools distributed evenly.

Storage of parts for protection and easy access is another important point to consider. For a small project with only a few parts, the organization of components may not make much of a difference. For a larger project, however, organization of components can have a considerable impact on successful completion of the project. Organization methods can be simple or complex, depending on the size of the workshop and the size of the project. For a relatively small shop, plastic boxes such as those shown in Figure 6.4 will suffice. Some general storage guidelines include the following:

- Group components into logical divisions—for example, passive devices, active devices, and hardware.
- Use storage boxes that are clear or translucent so the contents can be readily seen without opening the box.
- Use boxes that allow for storage of large devices, such as vacuum tubes.

Soldering

The first rule of soldering is to use only enough heat to get the job done [2]. For circuit board work, product manufacturers recommend using a 20 W to 35 W temperature-controlled soldering iron. For hand-wired terminal strips, a 35 W iron is usually sufficient.

Applying solder is one thing; removing it is quite another. There are a number of products for removing solder from PWBs and components, including

- **Desoldering braid**   Fine copper formed into a braided wire, impregnated with rosin flux, and rolled up in a coil for easy handling.
- **Suction devices**   Add-on hardware to a soldering iron that is applied to the joint while the solder is molten, using either a squeeze bulb or spring-loaded plunger to suck the solder out.
- **Solder/desolder station**   Specialized tools for PWB repair.

Solder Types

Reliable construction and repair requires the use of quality solder [2]. The best type for most equipment is 0.028-in. (22-gauge) solder with a rosin core. Small-sized solder strands melt fast and lose less heat, which allows better control over the amount of solder applied to the joint. The solder should be made of virgin tin and desilvered lead, and be free from impurities such as zinc, aluminum, iron, copper, and cadmium. Do not try and save money on solder. Solder with the lowest melting point and the type that yields the strongest bond is made with a ratio of 63 percent tin and 37 percent lead, but 60-tin/40-lead is nearly as good.

For certain jobs, some manufacturers recommend *silver solder.* Silver solder contains about 3 percent silver along with the lead and tin. Silver solder is used when soldering such components as ceramic capacitors, which have silver-palladium fired onto the conductive surfaces. If common tin/lead solder were used, it might absorb some of the silver from the component, causing a weak joint and poor adhesion. The small amount of silver in the solder reduces migration of the silver from the component connections. Silver solder is used in the same manner as ordinary solder and performs essentially the same, except for a slightly higher melting point. If the device literature calls for silver solder, use it.

Solder Flux

The rosin core of the solder is usually adequate to clean oxide from the joint as soldering takes place [2]. In some cases, however, additional flux may be required. It is good practice to have some high-quality liquid rosin flux on hand. Rosin is a nonconductive, noncorrosive flux that is recommended for work on electronic circuits. Rosin flux is, however, sticky and will collect dust if allowed to remain on the solder joint. After soldering, clean off any excess flux to remove a potential cause of future problems.

Moving Heat to the Joint

Most soldering irons are sold on the basis of heating element wattage [2]. This rating is, unfortunately, commonly the subject of misunderstandings. The wattage only indicates the potential heat an iron can produce. The amount of heat that actually reaches the tip will be considerably less than the iron's rated heat. The amount of heat delivered to the work point is determined by the

- Heat transfer efficiency of the iron
- Shape of the soldering tip
- Distance between the heating element and the work

In most electronics applications, 650°F is the minimum amount of heat required to *reflow* a solder joint. However, this does not take into consideration heat lost from the tip. A heat reserve also must be figured into the choice of an iron. But be careful: too much heat can ruin components and carbonize the flux before it has a chance to do its job. In general, most connections require about 800°F. Large connections, such as braided wire grounds or heavy-gauge wire, require about 1,000°F.

Soldering iron tips available today usually are made of copper with a thick iron or nickel plating for long life. They must never be *redressed;* filing or grinding will destroy the tip. Coated tips also need to be retinned less often than copper tips. Corrosion is the worst enemy of a soldering tip because it prevents efficient transfer of heat to the work point. When a clad tip becomes corroded, sand lightly with a piece of emery cloth and then retin. Never retin a soldering iron while it is hot. Let the iron cool, then warm for about one and a half minutes, and apply flux-core solder. This procedure is recommended because corrosion is faster at higher temperatures. Tinning while the tip is cooler will provide the best soldering surface. Replace a clad tip when corrosion has eaten through the plating.

A selection of various soldering tips is advisable, particularly for projects using PWBs. The optimal shape for a soldering tip for a circuit board is different than for a terminal strip. Tips are inexpensive; having several types on hand is a good practice.

The dangers of using too much heat to solder or desolder a component are obvious, but what about not using enough heat? If the solder connection is not heated sufficiently, a *rosin joint* can result. Although a rosin joint looks nearly identical to a good solder joint, the flux resins in a rosin joint insulate the component lead and prevent reliable contact with the PWB trace or terminal strip.

After the device has been soldered, inspect for rosin joints. A good solder joint is smooth and shiny; a rosin joint is dull gray and full of pin holes. As illustrated in Figure 6.5*a*, when both the component lead and the PWB foil are heated at the same time, the solder will flow onto the lead and the foil evenly, resulting in a good electrical connection between the lead and the foil. However, when the lead or the foil is not heated sufficiently, the solder will not flow properly and a poor electrical connection can result.

Techniques for soldering components and wires to terminal strips are similar to those for PWBs, except that the leads are typically held in place by crimping prior to soldering. Figure 6.5*b* shows the proper techniques.

Contact Cleaner/Lubricants

Chemicals are an important element in building or servicing electronic devices [2]. The right chemical applied to the right point will often help identify a problem or restore a system to proper operation. Every technician knows that periodic cleaning of electronic hardware is essential to reliability. Cleaning removes oxides, dust, grease, and other environmental contaminants that can affect electrical conductivity and trap heat. Cleaning also imparts a degree of protection against frictional wear, corrosion, and static build-up.

Chemicals for Troubleshooting

A number of solvents and cleaning agents are available in both aerosol and liquid form to remove performance-inhibiting contaminants from electronic equipment [2]. The principal applications of these chemicals include

- Degreasing contacts and connectors
- Removing organic flux after soldering

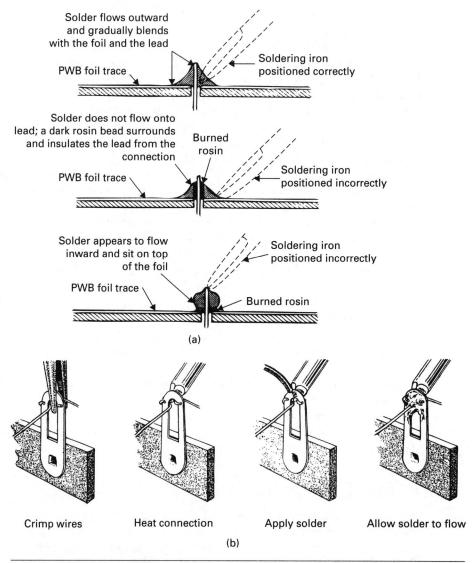

Solder flows outward
and gradually blends
with the foil and the lead

Soldering iron
positioned correctly

PWB foil trace

Solder does not flow onto
lead; a dark rosin bead surrounds
and insulates the lead from the
connection

Burned
rosin

Soldering iron
positioned incorrectly

PWB foil trace

Solder appears to flow
inward and sit on top
of the foil

Soldering iron
positioned incorrectly

PWB foil trace

Burned rosin

(a)

Crimp wires   Heat connection   Apply solder   Allow solder to flow

(b)

FIGURE 6.5   Soldering techniques: (*a*) PWB soldering techniques, (*b*) terminal strip soldering techniques. (*After* [3].)

Although personal preference and habit often determine which cleaning solvents are used in the shop, other, more objective factors, such as safety, effectiveness, and convenience, should be considered. Electronic cleaning solvents can be divided into four major categories based on chemical composition:

- Chlorofluorocarbons
- Chlorinated solvents
- Alcohols
- Blends of chemicals

Aerosols are often preferred because they deliver a continually fresh supply of uncontaminated solvent with sufficient pressure to dislodge and remove even encrusted grease without scrubbing. Aerosol solvents can be applied effectively using a lint-free cloth for catching overspray and wiping the PWB and/or component. Carbon-dioxide propelled aerosol systems provide the greatest initial spray pressure.

A number of excellent chemical blends are also available. Each blend possesses a specific measure of solvency, reactivity, and flammability. When using these products, check the package label for contents and capabilities, as well as precautions against adverse reactions with materials to be cleaned. Some common chemicals for electronics work are shown in Figure 6.6.

Applicators

For many years, household cotton swabs have been routinely used to apply solvents and remove contaminants from electronic hardware [2]. Although inexpensive and readily available, everyday cotton swabs produce lint and are not correctly sized or textured for many cleaning jobs. A superior alternative is found in the large variety of specialized swabs and applicators made with tips of polyurethane foam, polyester cloth, and other synthetic materials. These products are designed to be highly absorbent, thoroughly clean, and completely free of particles and extractables. Available in a wide assortment of sizes and shapes, foam-tip swabs are thermally mounted (without contaminating adhesives) on handles of various lengths and flexibility for a variety of cleaning jobs.

Presaturated Pads and Swabs

Disposable presaturated pads and swabs contain a premeasured amount of liquid solvent in combination with an application device—either a cloth pad or a foam-tip

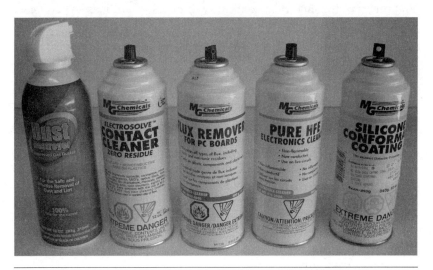

FIGURE 6.6   Common chemicals for a well-equipped shop

swab—sealed in individual foil packets [2]. The user simply tears open a packet, cleans the surfaces, and then discards the used cleaning item. Premoistened pads and swabs are easy to store and use. They are used only once, preventing exposure of sensitive surfaces to contamination. Premoistened pads are highly recommended for cleaning PWB connectors and equipment housings. Presaturated nonresidual isopropanol swabs are especially useful for hard-to-reach areas.

Compressed Gas Dusters

Made from a microscopically clean, moisture-free inert fluorocarbon (dichlorodifluoromethane), compressed gas dusters deliver powerful jet action to remove dust and particulates from even the most inaccessible areas of electronic equipment [2]. Compressed gas dusters are useful for cleaning delicate components in applications where liquid solvents are inappropriate or inconvenient. Any surface, no matter how sensitive, can be safely cleaned with a compressed gas duster without risk of abrasive damage or contamination.

Exercise care, however, when using compressed gas duster products. Make sure the stream of air is used to move dust and dirt out of the equipment being serviced, not simply to another part of the hardware. It is of little use to move dirt from where you can see it to where you can't see it. The contaminants may still cause problems for the equipment, regardless of whether they are visible.

Contact Cleaner/Lubricants

Chemical products that combine precision cleaning agents with fine-grade lubricants provide a convenient method to clean, restore, and protect electronic contacts [2]. Available as aerosol sprays and premoistened pads, contact cleaner/lubricants are especially useful for protecting thin metal surfaces against frictional wear, oxidation, and corrosion. Applied periodically to contacts, cleaner/lubricants prolong the life and preserve the electrical continuity of metallic surfaces. When choosing a product of this type, look for the following properties:

- High temperature lubricity
- Low volatility
- Controlled *creep* onto neighboring surfaces
- Oxidation stability (to prevent the formation of gummy by-products)
- Compatibility with adjacent materials

When using these products, take steps to ensure that the cleaner does not spray into or drip onto other components. Place an absorbent pad or cloth adjacent to the application point to catch any overspray. Use only the amount of cleaner necessary to reach the contacts. Excessive application will result in run-off of the cleaner onto the chassis.

Keep in mind that contact cleaners can do only so much to restore switches, potentiometers, and other components. At some point, replacement of the device may be the best solution.

Test Equipment and Troubleshooting

The failure modes that may be experienced with an electronic system are widely varied [5]. Some general rules, however, may be applied to nearly all troubleshooting efforts. First, follow the natural signal flow as it moves through the system. When a failure node is found, trace backward through the circuit until a point is located where the input is as expected but the output is not as expected. To accomplish this requires good documentation of the system and the proper test equipment. Table 6.3 lists the test equipment that can be found in a well-equipped shop.

Some of the equipment listed in Table 6.3 can be realized as software programs running on a computer, using either the built-in sound card or an external (usually USB) interface device. Such systems provide considerable flexibility and permit easy documentation of test measurements.

A well-equipped workbench intended for audio work is shown in Figure 6.7.

Digital Multimeter

The most basic tool in today's arsenal of electronic test instruments is the digital multimeter (DMM) [2]. Beyond the basic voltage-resistance-current measurements, DMMs provide a number of useful features and functions, including

- **Audible continuity check**   The instrument emits a beep when the measured resistance value is less than 5 Ω or so. This permits point-to-point continuity tests to be performed without actually looking at the DMM.

TABLE 6.3   Basic and Optional Test Equipment for Audio Work

Basic Test Equipment	AC voltmeter
	Audio signal generator
	Bench digital multimeter (DMM)
	Harmonic distortion analyzer
	Oscilloscope (dual-trace preferred, storage optional)
Optional Test Equipment	AM/FM alignment generator (for work on receivers)
	Audio frequency sweep generator
	Capacitance substitution box
	Frequency counter
	High-voltage power supply (0 to 500 Vdc)
	Intermodulation distortion analyzer
	Low-voltage power supply (0 to 50 Vdc)
	Oscilloscope calibrator
	Resistance substitution box
	Volt-ohm meter (VOM) with analog meter movement

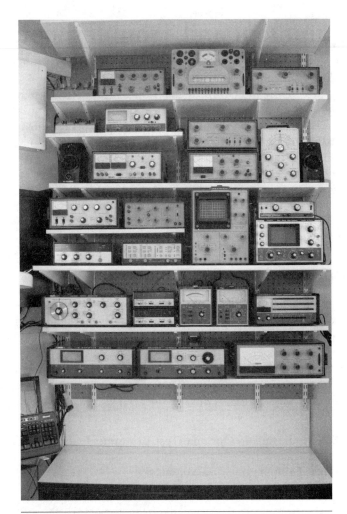

FIGURE 6.7   Test equipment for audio project construction and troubleshooting. (Observant readers will note that all of the test equipment in this shop is classic 1970s era Heathkit instruments, which seems appropriate for building classic tube amplifiers.)

- **Audible high-voltage warning**   The instrument emits a characteristic beep when the measured voltage is dangerously high (100 V or more). This safety feature alerts the user when making measurements.
- **Sample-and-hold**   The instrument measures and stores the sampled value after it has stabilized. Completion of the sample function is signaled by an audible beep. This feature permits the user to concentrate on probe placement, not on trying to read the instrument.

- **Continuously variable tone** The instrument produces a tone that varies in frequency with the measured parameter. This feature permits a circuit to be adjusted for a peak or null value without actually viewing the DMM. It also may be used to check for intermittent continuity.
- **Input overload hardening** Protection devices are placed at the input circuitry to prevent damage in the event of an accidental overvoltage. Fuses and semiconductor-based protection components are used.
- **Automatic shut-off** A built-in timer removes power to the DMM after a preset period of nonoperation. This feature extends the life of the internal batteries.

DMM functions are further expanded by the wide variety of probes and accessories available for many instruments. Specialized probes include

- **High-voltage** Used for measuring voltages above 500 V; many probes permit voltages of 50 kV or more to be measured.
- **Clamp-type current** Permits current in a conductor to be measured without breaking the circuit. Probes are commonly available in current ranges of 1–2 A and up.
- **Demodulator** Converts a radio frequency signal into a DC voltage for display on the DMM.
- **Temperature** Permits the DMM to be used for display of temperature of a probe device.

Specifications

The ultimate performance of a DMM is determined by its inherent accuracy—how closely the meter indicates the actual value of a measured signal, specified in percent of error [2]. Zero percent indicates a perfect meter. Key specifications include

- **Frequency response** The range of frequencies that can be measured by the meter without exceeding a specified amount of error.
- **Input impedance** The combined AC and DC resistance at the input terminals of the multimeter. The impedance, if too low, can load sensitive circuits and result in measurement errors. An input impedance of 1 MΩ or greater will prevent loading.
- **Precision** The degree to which a measurement is carried out. As a rule, the more digits a DMM can display, the more precise the measurement will be.

AC Voltage Measurement

When a DC voltage is applied to a resistor, it produces a measurable temperature increase [4]. If an AC voltage is applied to the same resistor and produces the same temperature increase, then the AC voltage must be producing the same amount of power. Because this power produced by the AC voltage is averaged over a period of time, it is called "mean" (average) power. The AC voltage that produces the power is proportional to the square root of the mean power, and is called the root-mean-square (rms) voltage. AC meters are usually calibrated in rms voltage. For a sine wave, the most common AC voltage waveform, the rms value for each half-cycle is 0.707 times the peak waveform. The following relationships exist for sine-wave AC voltages:

- rms voltage = peak voltage × 0.707
- peak voltage = rms × 1.414
- peak-to-peak voltage = rms × 2.828
- rms voltage = peak-to-peak voltage × 0.3535

Signal Generator

In order to characterize the performance of an audio amplifier, it is necessary to observe how input signals are modified during the process of amplification. A perfect amplifier would provide an exact duplicate of the input signal at the output terminals that is increased in level by the desired amount. Perfect amplifiers do not exist in the real world, and so it is important to have some manner of determining the extent to which the signal is modified by the amplifier. While critical listening in a controlled environment is the ultimate performance test for an amplifier, such tests are not always repeatable from one observer to another and the documentation is subjective.

For the foregoing reasons, initial tests on almost all audio devices begin with a signal generator of known characteristics. Audio signal generators typically have a frequency range of 10 Hz (or less) to 100 kHz (or more), adjustable in steps and/or continuously variable increments. Low-distortion sine waves are provided, typically exhibiting less than 0.3 percent harmonic distortion. In addition to the sine wave, a separate square wave output is often provided, covering the same frequency span. The output voltage is usually adjustable up to 10 V AC in a series of steps.

A sweep generator performs the same functions as an audio frequency generator, with the added feature that the output frequency can be made to change (sweep) from a set low frequency point to a set high frequency point over a specified period. Such devices are quite useful for observing the frequency response of an audio amplifier, where—for example—the input signal sweep begins at 20 Hz and ends at 20 kHz and the output is observed on an oscilloscope. Such measurements may not eliminate the need for tests at discrete frequencies, but they do quickly characterize the system. This is particularly helpful for optimizing a circuit design, for example.

The sweep generator typically provides sine, square, and triangle wave outputs at voltages up to 10 V AC in a series of steps. The start and stop points for the frequency sweep can be set individually, and the speed of the sweep is usually adjustable over a wide range. The sweep can be set to occur once, cycle repeatedly, or trigger on some preset event.

Frequency Counter

A frequency counter provides an accurate measure of signal cycles or pulses over a standard period [2]. It is used to totalize or to measure frequency, period, frequency ratio, and time intervals. Key specifications for a counter include

- **Frequency range**   The maximum frequency that the counter can resolve.
- **Resolution**   The smallest increment of change that can be displayed. The degree of resolution is selectable on many counters. Higher resolution usually requires a longer acquisition time.
- **Sensitivity**   The lowest amplitude signal that the instrument will count (measured in fractions of a volt).
- **Timebase accuracy**   A measure of the stability of the timebase. Stability is measured in parts per million (ppm) while the instrument is subjected to temperature and operating voltage variations.

Signal Tracer

An audio signal tracer is a convenient instrument for go/no testing and troubleshooting of audio or radio circuits. The instrument can be as simple as an audio amplifier, perhaps with an integral voltmeter, whose input can be connected at various points in a system under test to check for the presence or absence of a signal. The signal tracer is not intended to provide a qualitative measure of the signal characteristics, except in a general way (e.g., excessive hum), but rather to quickly locate the point at which an expected signal is lost.

The signal tracer may also include a detector (typically in the test probe) to facilitate troubleshooting radio circuits.

Distortion Analyzer

There are two basic types of distortion analyzers for audio applications: *harmonic distortion* and *intermodulation distortion* [6]. Instruments for measuring harmonic distortion are more common, although both are used to characterize the performance of audio systems. Both devices utilize a signal source of known characteristics that is passed through a circuit (or device) under test and the differences in the output signal (other than amplitude) are measured.

Harmonic distortion is measured at discrete frequencies using an audio frequency sine wave generator connected to the input of the unit under test and the harmonic distortion analyzer connected to the output. The instrument is adjusted to remove the input sine wave using a variable notch filter. With the fundamental frequency removed, the components that remain are harmonics of the input signal. By measuring the amplitude of the harmonics, the *total harmonic distortion* (THD) can be found (see Figure 6.8.) Harmonic distortion measured in this manner is actually a measure of THD plus whatever noise is present at the output from all sources. As such, the more proper term is THD + noise. For a low-noise amplifier, the noise component can be largely ignored insofar as THD measurement is concerned. Unless otherwise specified, THD usually refers to THD + noise.

For more detailed analysis of THD + noise, the harmonic distortion instrument usually provides an output of the residual signal (THD + noise), which can be displayed on an oscilloscope or spectrum analyzer. Through proper adjustment of the display, the harmonics can be examined separately from the noise.

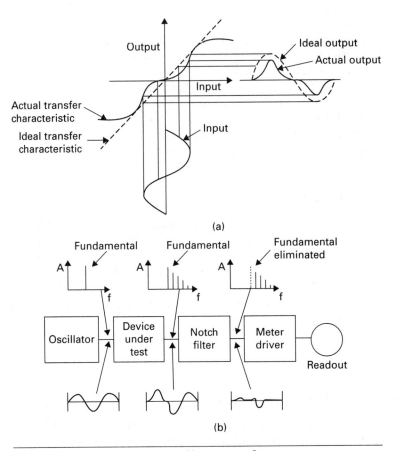

FIGURE 6.8 Measurement of harmonic distortion: (*a*) transfer characteristic of a typical device; (*b*) block diagram of a distortion analyzer. (*After* [6].)

Intermodulation distortion (IMD) measurement begins with the input of two signals of known characteristics (frequency and relative amplitude) that are sampled at the output of the device under test. One signal is set to be a low frequency (e.g., 60 Hz) and the other a high frequency (e.g., 7 kHz). The IMD instrument measures the amount to which the two signals interact. Nonlinearities in a circuit can cause it to behave like a mixer, generating the sum and difference frequencies of two signals that are present. These same imperfections generate harmonics of the signals, which then can be mixed with other fundamental or harmonic frequencies. The output signal is examined for modulation of the upper frequency by the low-frequency tone. The amount by which the low-frequency tone modulates the high-frequency tone indicates the degree of nonlinearity (see Figure 6.9.) As with harmonic-distortion measurement, this test is typically performed with a dedicated distortion-analysis instrument (see Figure 6.10).

While other options are available, THD and IMD are usually specified in percent distortion, relative to the signal level at the input of the instrument.

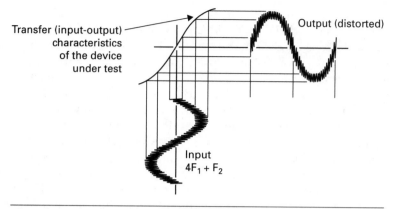

FIGURE 6.9 Intermodulation distortion test of transfer characteristic. (*After* [6].)

Oscilloscope

The oscilloscope is one of the most general-purpose of all test instruments [2]. A scope can be used to measure voltages, examine waveshapes, check phase relationships, and countless other functions. In addition to familiar choices such as bandwidth and risetime, and single or multitrace operation, oscilloscopes come in stand-alone and hardware/software versions. For the latter case an input device of some type is connected to a computer running the appropriate software.

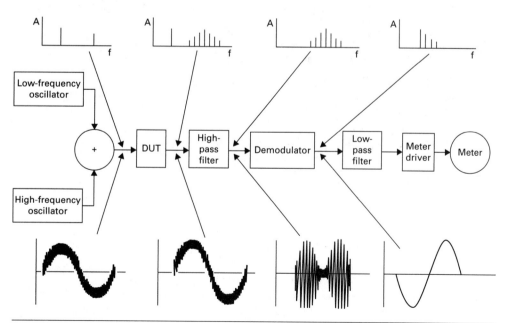

FIGURE 6.10 Block diagram of an intermodulation distortion analyzer measuring the performance of a device under test (DUT). (*After* [6].)

A number of parameters are used to characterize the performance of an oscilloscope. Key parameters include bandwidth and risetime.

Oscilloscope bandwidth is defined as the frequency at which a sinusoidal signal will be attenuated by a factor of 0.707 (or reduced to 70.7 percent of its maximum value). This is referred to as the –3 dB point. Bandwidth considerations for sine waves are straightforward. For square waves and other complex waveforms, however, bandwidth considerations become substantially more involved.

Square waves are made up of an infinite number of sine waves: the fundamental frequency plus mostly odd harmonics. Fourier analysis shows that a square wave consists of the fundamental sine wave plus sine waves that are odd multiples of the fundamental. Figure 6.11 illustrates the mechanisms involved. The fundamental frequency contributes about 81.7 percent of the square wave. The third harmonic contributes about 9.02 percent, and the fifth harmonic about 3.24 percent. Higher harmonics contribute less to the shape of the square wave. Because a square wave signal includes these harmonic components, it is useful for quickly characterizing the performance of a circuit over a broad range of frequencies.

The risetime performance required of an oscilloscope depends on the degree of accuracy needed in measuring the input signals. A risetime accuracy of within 2 to 3 percent can typically be obtained from an instrument that is specified to have approximately five times the risetime performance of the signal being measured.

Making Measurements

The oscilloscope is a measurement instrument, displaying voltage waveforms within powered circuits [2]. To conduct certain types of waveform analysis, a separate signal generator is required to inject test signals of known characteristics. With a conventional

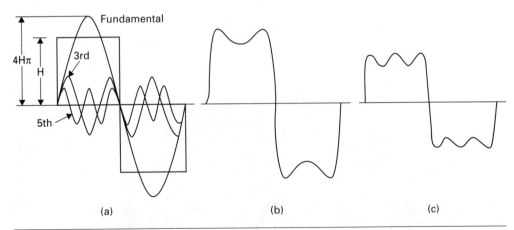

FIGURE 6.11   The mechanisms involved in square waves: (*a*) individual waveforms involved; (*b*) waveshape resulting from combining the fundamental and the first odd harmonic (the third harmonic); (*c*) waveshape resulting from combining the fundamental and the first two odd harmonics (third and fifth harmonics). (*After* [7].)

oscilloscope, actual measurements of the waveform are done visually by the operator. The graticule divisions along the trace are counted and multiplied by the timebase setting to determine the time interval. The TIME/DIV control determines the value of each increment on the X-axis of the display. To determine voltage, the operator must estimate the height of a portion of the trace and multiply it by the voltage range selected on the instrument. The VOLT/DIV control determines the value of each increment on the Y-axis of the display. On a dual channel scope, two controls are provided, one for each source.

The need for interpolation introduces potential inaccuracies in the measurement process. Determination of exact values is especially difficult in the microsecond ranges, where even small increments can make a big difference in the measurement. It is also difficult with a conventional scope to make a close-up examination of specific portions of a waveform.

The digital storage oscilloscope (DSO) offers a number of significant advantages beyond the capabilities of conventional instruments. A DSO can store in memory the signal being observed, permitting in-depth analysis that is otherwise very difficult. Because the waveform resides in memory, the data associated with the waveform can be stored for processing or recalled at a later time.

Computer-based oscilloscopes offer features similar to a DSO. The capabilities of such systems are largely determined by the interface device used (typically, a purpose-built add-on) and the processing power of the computer itself.

Tube Tester

If new tube-testing instruments are being built today, they are rare; instruments built in the 1970s and earlier are usually the devices available. A vacuum tube possesses a number of operating characteristics, any one of which may be used to indicate—to a certain degree—the operational capabilities of a given device. Several techniques are available to measure the performance of a tube, each one subject to certain limitations. A tube tester cannot provide a complete and accurate characterization of the performance of a given tube under all possible operating conditions. These instruments are, however, quite useful for go/no-go testing and basic troubleshooting. In the final analysis, the only test instrument that really matters is the amplifier or receiver in which the tube is used. Still, when confronted with equipment that is not working correctly, the tube tester can be useful.

A number of common techniques are used in tube testers, including the following [8]:

- *Emissions testing,* where the emission capabilities of the cathode are measured to determine the overall quality of the vacuum tube. This can be accomplished by connecting all of the grids to the plate and operating the tube as a rectifier. The actual emission of the cathode is then compared to a predetermined value accepted as nominal for that tube type. If the cathode should have one particularly active portion, the emission checker will indicate the quality of the tube as "good," even though the remainder of the cathode may be inactive. On the other hand, most coated cathodes are capable of large emission, often far in excess of the emission required for a particular application.

- *Transconductance testing,* where a predetermined voltage is placed on each tube element, creating a plate current flow. Measurement of this current indicates the transconductance of the device under static conditions.
- *Dynamic transconductance testing,* which improves on the transconductance testing approach described earlier by adding an input signal source. By applying a signal to the device under test, the action of the plate current will be similar to that experienced in the actual application, varying in relationship to the input. Although this technique gives an indication of how the device will operate under signal conditions, it is still limited in scope. Certain types of tubes cannot be satisfactorily checked under any conditions other than full power, such as vertical or horizontal output tubes used in television receivers.
- *Power output testing,* where actual operating conditions are simulated by the instrument. Because the input and output powers are known, the true capabilities of the device can be determined. This technique is best suited to testing tubes used in power amplifiers, where output capability is a major concern.

The selection of which of these techniques—or combination of techniques—is used in any given instrument is a function of complexity and cost.

Power Supplies

A bench power supply can be a valuable tool in equipment development and/or troubleshooting. A wide variety of supplies are available to meet particular needs. For work with vacuum tube circuits, the following capabilities are useful:

- AC power for filaments, typically 6.3 V AC and 12.6 V AC
- Grid bias supply adjustable from 0 V to –100 V DC at 1 mA or so
- Plate supply adjustable from 0 V to +400 V DC at 100 mA or so

Additional features useful for work with solid-state systems include

- Bipolar adjustable 0 V to 20 V DC outputs, independent or tracking, at 0.5 A or so for each output
- Fixed +5 V DC output at 2 A or more
- Adjustable 0 V to +50 V output at 1 A or so

In addition, a line voltage power supply (120 V AC) that is adjustable over a small range (such as ±20 percent) is useful for measuring how a system reacts to changes in the input line voltage.

Troubleshooting Guidelines

When presented with a problem, proceed in an orderly manner to trace it down [2]. Many failures are simple to repair if you stop and think about what's happening. Looking over the schematic diagram for 10 to 15 minutes to consider the possible causes can save hours of trial-and-error troubleshooting. Never rush through a troubleshooting job.

When checking inside the unit, look for changes in the physical appearance of components. An overheated resistor or leaky electrolytic capacitor may be the cause of the problem, or point to the cause. Devices never fail without a reason. Try and piece together the sequence of events that led to the problem. Then, the cause of the failure—not just the more obvious symptoms—will be corrected. When working with direct-coupled transistors, a failure in one device may cause a failure in another, so check all semiconductors associated with one found to be defective. In high-voltage power supplies, look for signs of loose connections and overheating. Do not overlook the possibility of tube failure. Tubes can fail in unusual ways, and substitution may be the only practical test. Troubleshooting through the process of elimination involves isolating various portions of the circuit—one section at a time—until the defective component is found.

Never touch anything inside the unit without first removing all AC power and discharging all filter capacitors with a test lead to ground. Most plate power supplies include bleeder resistors to drain off the charge on the capacitor(s) in the circuit. However, some designs do not include this safety feature. Be careful.

In the event that a wire is disconnected to isolate a circuit, temporarily wrap it with electrical tape and secure the connector so it will not arc over to ground or another component if power is applied. Analyze each planned test before it is conducted. Every test in the troubleshooting process requires time, and so steps should be arranged to provide the greatest amount of information about the problem.

Safety Considerations

Electrical safety is important when working with any type of electronic hardware. Because vacuum tubes operate at high voltages and currents, safety is doubly important. The primary areas of concern, from a safety standpoint, include electric shock and hot surfaces of vacuum tube devices.

Electric Shock

Surprisingly little current is required to injure a person [2]. Studies at Underwriters Laboratories (UL) show that the electrical resistance of the human body varies with the amount of moisture on the skin, the muscular structure of the body, and the applied voltage. The typical hand-to-hand resistance ranges between 500 Ω and 600 kΩ, depending on the conditions. Higher voltages have the capability to break down the outer layers of the skin, which can reduce the overall resistance value. UL uses the lower value, 500 Ω, as the standard resistance between major extremities, such as from the hand to the foot. This value is generally considered the minimum that would be encountered and, in fact, may not be unusual because wet conditions or a cut or other break in the skin significantly reduces human body resistance.

Effects on the Human Body

Table 6.4 lists some effects that typically result when a person is connected across a current source with a hand-to-hand resistance of 2.4 kΩ [2]. The table shows that a current of approximately 50 mA will flow between the hands if one hand is in contact with a 120 V AC source and the other hand is grounded. The table indicates that even the relatively small current of 50 mA can produce ventricular fibrillation of the heart, and perhaps death. Medical literature describes ventricular fibrillation as rapid, uncoordinated contractions of the ventricles of the heart, resulting in loss of synchronization between heartbeat and pulse beat. Unfortunately, once ventricular fibrillation occurs, it will continue. Barring resuscitation techniques, death can ensue within a few minutes.

The route the current takes through the body has a significant effect on the degree of injury. Even a small current passing from one extremity through the heart to another extremity is dangerous and capable of causing severe injury or electrocution. There are cases where a person has contacted extremely high current levels and lived to tell about it; however, usually when this happens, the current passes only through a single limb and not through the body.

Current is not the only factor in electrocution. Figure 6.12 summarizes the relationship between current and time on the human body. The graph shows that 100 mA flowing through a human adult body for 2 s will cause death by electrocution. An important factor in electrocution, the *let-go* range, also is shown on the graph. This range is described as the amount of current that causes "freezing," or the inability to let go of the conductor. At 10 mA, 2.5 percent of the population will be unable to let go of a "live" conductor. At 15 mA, 50 percent of the population will be unable to let go of an energized conductor. It is apparent from the graph that even a small amount of current can "freeze" someone to a conductor. Table 6.5 lists required precautions for persons working around high voltages.

TABLE 6.4 The Effects of Current on the Human Body

1 mA or less	No sensation, not felt
More than 3 mA	Painful shock
More than 10 mA	Local muscle contractions, sufficient to cause "freezing" to the circuit for 2.5 percent of the population
More than 15 mA	Local muscle contractions, sufficient to cause "freezing" to the circuit for 50 percent of the population
More than 30 mA	Breathing is difficult, can cause unconsciousness
50 mA to 100 mA	Possible ventricular fibrillation of the heart
100 mA to 200 mA	Certain ventricular fibrillation of the heart
More than 200 mA	Severe burns and muscular contractions; heart more apt to stop than to go into fibrillation
More than a few amperes	Irreparable damage to body tissues

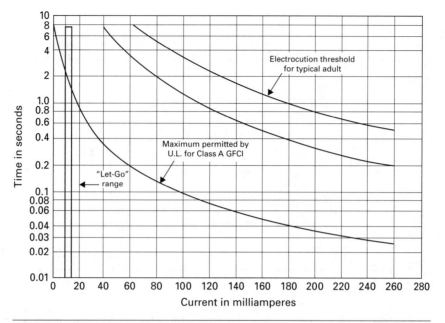

FIGURE 6.12 Effects of electric current and time on the human body. Note the "let-go" range. (*From* [2]. *Used with permission.*)

Circuit Protection Hardware

The typical primary panel or equipment circuit breaker or fuse will not protect a person from electrocution [2]. In the time it takes a fuse or circuit breaker to blow, someone could die. However, there are protection devices that, properly used, help prevent electrocution. The ground-fault current interrupter (GFCI), shown in Figure 6.13, works by monitoring the current being applied to the load. The GFCI uses a differential

TABLE 6.5 Safety Practices for Working Around High-Voltage Equipment

Remove all AC power from the equipment. Do not rely on internal switches to remove dangerous AC.

Discharge all capacitors using a grounded probe.

Do not remove, short circuit, or tamper with interlock switches on access covers, doors, or enclosures.

Keep away from live circuits.

Allow any component to completely cool down before attempting to replace it.

If a leak or bulge is found on the case of an electrolytic capacitor, do not attempt to service the part until it has completely cooled.

Avoid contact with hot surfaces within the system.

Know your equipment and do not take chances.

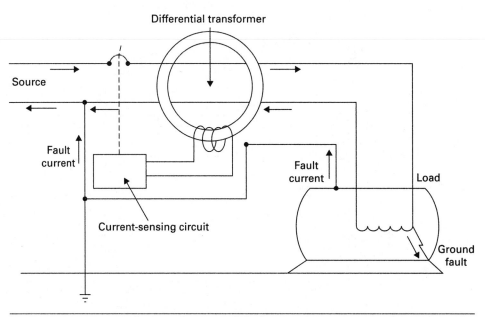

FIGURE 6.13 Basic design of a ground-fault current interrupter (GFCI)

transformer and looks for an imbalance in load current. If a current (5 mA, ±1 mA) begins to flow between the neutral and ground or between the hot and ground leads, the differential transformer detects the leakage and opens up the primary circuit within approximately 2.5 ms.

GFCIs will not protect a person from every type of electrocution. If the victim becomes connected to both the neutral and the hot wire, the GFCI may not detect an imbalance.

When working around high voltages, always look for grounded surfaces. Keep hands, feet, and other parts of the body away from any grounded surface. Even concrete can act as a ground if the voltage is sufficiently high. Of course, the best troubleshooting methodology is never to work on any circuit without being certain that no hazardous voltages are present.

Be certain that any equipment on the test bench is fed with AC power provided through an outlet protected by a GFCI device.

First Aid Procedures

All persons working around high-voltage equipment should be familiar with first aid treatment for electric shock and burns [2]. Always keep a first aid kit on hand. Obtain detailed information from the local heart association or Red Cross chapter. Personalized instruction on first aid usually is available locally.

Protective Eyewear

It is advisable to wear glasses or other protective eyewear when building or working on electronic equipment. Injuries can result from solder splashes, cleaning chemicals, and other unexpected events. When trimming the leads of a component, be careful to hold the excess length so that when cut the lead does not fly toward the face.

Parts List

It is beyond the scope of this book to detail all of the bulk supplies and parts necessary to construct a given vacuum tube project; having said that, Table 6.6 provides a starting point for some of the more obvious items. Note that for each item a manufacturer and

TABLE 6.6   Miscellaneous Bulk Supplies and Parts

Description	Quantity	Manufacturer	Part No.	Allied Stock No.	Notes
Hookup wire, #18 solid, black, 300 V	100 ft	Alpha Wire	3055/1 BK005	696-9302	AC power hot
Hookup wire, #18 solid, green, 300 V	100 ft	Alpha Wire	3055/1 GR005	696-9304	AC power ground
Hookup wire, #18 solid, white, 300 V	100 ft	Alpha Wire	1565 WH005	696-0041	AC power neutral
Hookup wire, #18 solid, yellow, 300 V	100 ft	Alpha Wire	3055/1 YL005	696-9305	Output/speaker
Hookup wire, #18 solid, red, 300 V	100 ft	Alpha Wire	3055/1 RD005	696-9303	Power control functions (if used)
Hookup wire, #22 solid, yellow, 300 V	100 ft	Alpha Wire	3051/1 YL005	696-9285	Input signal lines
Hookup wire, #22 solid, black, 300 V	100 ft	Alpha Wire	3051/1 BK005	696-9282	Ground
Hookup wire, #22 solid, red, 600 V	100 ft	Alpha Wire	2855/1 RD005	696-1223	B+ power supply
Hookup wire, #22 solid, white, 300 V	100 ft	Alpha Wire	3051/1 WH005	696-9281	B– power supply
Hookup wire, #22 solid, green, 300 V	100 ft	Alpha Wire	3051/1 GR005	696-9284	Heater power supply
Hookup wire, #22 solid, blue, 300 V	100 ft	Alpha Wire	3051/1 BL005	696-9286	Ancillary circuits (if used)

(continued)

TABLE 6.6 Miscellaneous Bulk Supplies and Parts (*Continued*)

Description	Quantity	Manufacturer	Part No.	Allied Stock No.	Notes
Shielded audio cable, #24 stranded, pair	100 ft	Belden	8641 060100	216-1102	Audio interconnections
3 conductor shielded audio cable, #24 stranded	100 ft	Belden	9533 060100	216-2810	Audio interconnections
Expandable braided sleeving	100 ft	Alpha Wire	GRP1101/2 BK005	708-4792	
Heat shrink tubing	100 ft	Alpha Wire	FITKIT 221BK BK032	708-0442	Stock number for assorted sizes
Cable tie, 4 in.	100	3M	06200	617-9402	
Wire-nut connector	100	Ideal	30-073J		Twist-on connectors for 22 to 14 AWG wire
6-32 screw, Phillips pan head, stainless steel, 1/2-in	100	Bolt Depot[1]	1336		
6-32 screw, Phillips pan head, stainless steel, 3/4-in	100	Bolt Depot	1338		
6-32 screw, Phillips pan head, stainless steel, 1-in	100	Bolt Depot	1339		
6-32 screw, Phillips flat head, stainless steel, 1/2-in	100	Bolt Depot	1238		
6-32 nut, stainless steel, captive lockwasher	400	Bolt Depot	12019		
#6 washer, stainless steel	300	Bolt Depot	2942		
4-40 screw, Phillips, pan head, stainless steel, 3/8-in	100	Bolt Depot	9628		
4-40 screw, Phillips, flat head, stainless steel, 3/8-in	100	Bolt Depot	3717		

(*continued*)

TABLE 6.6   Miscellaneous Bulk Supplies and Parts (*Continued*)

Description	Quantity	Manufacturer	Part No.	Allied Stock No.	Notes
4-40 screw, Phillips pan head, stainless steel, 5/8-in	100	Bolt Depot	5313		
4-40 screw, Phillips flat head, stainless steel, 5/8-in	100	Bolt Depot	5303		
#4 washer, stainless steel	100	Bolt Depot	5563		
#4 internal tooth lockwasher, stainless steel	100	Bolt Depot	4087		
#4 nut, stainless steel, captive lockwasher	100	Bolt Depot	12018		
Threaded standoff	100	Keystone	1895	839-0781	
Grommet	100	H. H. Smith	2174	920-2174	
Cable clamp	100	H. H. Smith	8943	920-8943	
Ground lug	100	Keystone	904	839-2390	
Paint, finish, black	1	Rust-Oleum	245217		Black hammered

[1] Bolt Depot: www.boltdepot.com

part number are listed. In addition, a stock number[8] is provided for online ordering, if you so choose.

In nearly all cases, the part numbers given in the table represent bulk amounts, such as 100 count, or 100 ft. Unless you will be constructing a large number of projects, these quantities will be excessive. The intent here is to provide specific examples of hardware and other supplies so the reader can acquire material in an efficient manner.

Note that the hookup wire specified in Table 6.6 includes designated functions for each wire size and color. These color-code assignments are essentially arbitrary; the reader can use any color for any purpose; still, color coding helps to simplify project assembly and maintenance. Readers are encouraged to choose some type of color-code scheme and document it appropriately.

Note also the voltage rating for each type of wire. All of the hookup wire specified in Table 6.6 is rated for 300 V except for the B+ wire (#22 solid, red), which is rated

---

[8] Allied Electronics, www.alliedelect.com.

for 600 V. While a margin of safety is typically built into insulation specifications, do not use hookup wire in a circuit where voltages greater than the specification will be present, as arcing can occur. Remember that rectifier tubes often utilize a combined heater/cathode. As such, the rating for wire used for the heater circuit must also be capable of handling the full B+ supply voltage.

Chapter 7

Tube Characteristics

This chapter provides specifications for the vacuum tubes used in the projects that follow. The data given here has been abstracted from the *RCA Receiving Tube Manual* [1]. For complete details on each tube type, see [1] or an equivalent publication. The devices are listed in numerical order.

5BC3A

The 5BC3A is used in power-supply applications for a variety of equipment types. Device pinout for the 9-contact novar base is shown in Figure 7.1. Vertical operation is preferred, but the tube may be operated in a horizontal position if pins 2 and 7 are in the vertical plane. It is important that this tube, like other power-handling devices, be adequately ventilated. Basic operating characteristics are charted in Figure 7.2. Selected tube parameters are listed in Table 7.1.

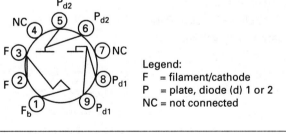

Legend:
F   = filament/cathode
P   = plate, diode (d) 1 or 2
NC = not connected

FIGURE 7.1   Terminal pinout of the 5BC3A. (*After* [1].)

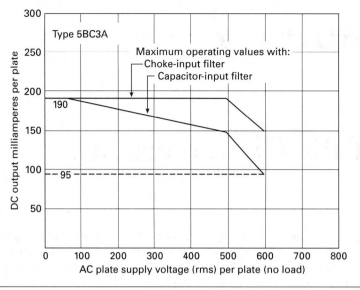

FIGURE 7.2 Characteristic curves of the 5BC3A. (*After* [1].)

TABLE 7.1 Key Parameters of the 5BC3A. (*After* [1].)

General Characteristics			
Heater voltage			5 V
Heater current			3A
Full-Wave Rectifier, Maximum Ratings (Design-Maximum Values)			
Peak inverse plate voltage			1700 V
Peak plate current (per plate)			1 A
AC plate voltage supply (per plate, rms)			Figure 7.2
Average output current (per plate)			Figure 7.2
Typical Operation, Choke Input to Filter			
AC plate-to-plate supply voltage		900 V	1100 V
Filter input choke		10 H	10 H
DC output voltage at input to filter (approx.)	Load current = 348 mA	340 V	–
	Load current = 275 mA	–	440 V
	Load current = 174 mA	355 V	–
	Load current = 137.5 mA	–	445 V

6EU7

The 6EU7 is a miniature type used in high-gain, resistance-coupled, low-level audio amplifier applications where low-hum and nonmicrophonic characteristics are important [1]. The nine-contact pinout is given in Figure 7.3. Device characteristic curves are shown in Figure 7.4. Selected tube parameters are listed in Table 7.2.

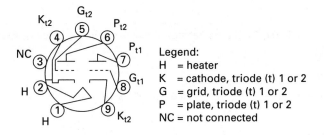

FIGURE 7.3   Terminal pinout of the 6EU7 tube. (*After* [1].)

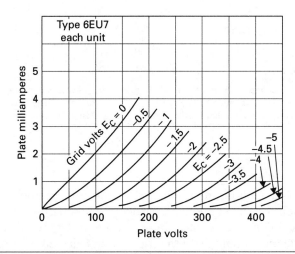

FIGURE 7.4   Characteristic curves for the 6EU7 tube. (*After* [1].)

TABLE 7.2   Key Parameters of the 6EU7. (*After* [1].)

General Characteristics

Heater (AC/DC)		6.3 V, 0.3 A
Peak heater-cathode voltage	Peak value	±200 V maximum
	Average value	±100 V maximum
Direct interelectrode capacitance (approx.), each unit	Grid to plate	0.15 pF
	Grid to cathode and heater	1.6 pF
	Plate to cathode and heater	0.2 pF
Equivalent noise and hum voltage (referenced to grid, each unit), average value[1]		1.8 μV rms

(continued)

TABLE 7.2   Key Parameters of the 6EU7. (*After* [1].) (*Continued*)

Class A$_1$ Amplifier, Maximum Ratings (Design-Maximum Values), Each Unit		
Plate voltage		330 V
Plate dissipation		1.2 W
Grid voltage	Negative-bias value	55 V
	Positive-bias value	0 V
Characteristics		
Plate voltage	100 V	250 V
Grid voltage	−1 V	−2 V
Amplification factor	100	100
Plate resistance (approx.)	80 kΩ	62.5 kΩ
Transconductance	1250 μmhos	1600 μmhos
Plate current	0.5 mA	1.2 mA

[1] Measured in "true rms" units under the following conditions: heater volts (AC) = 6.3 V; center tap of heater transformer connected to ground; plate supply = 250 V; plate load resistor = 100 kΩ; cathode resistor = 2700 Ω, cathode bypass capacitor = 100 μF; grid resistor = 0 Ω; and amplifier covering frequency range 25 Hz to 10 kHz.

6U8A

The 6U8A is a medium-mu triode and sharp-cutoff pentode tube well suited to use in high-fidelity audio equipment, particularly in phase-splitter and high-voltage gain applications. The nine-contact pinout is given in Figure 7.5. In phase-splitter circuits, the pentode unit should drive the triode unit. Selected tube parameters are listed in Table 7.3.

For certain voltage amplifier types, this device included, the maximum permissible screen grid (grid #2) input varies with the screen grid voltage, as shown in Figure 7.6 [1]. Full-rated screen grid input is permissible at screen grid voltages up to 50 percent of

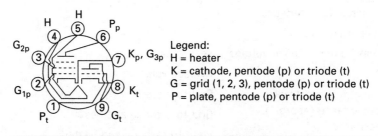

FIGURE 7.5   Terminal pinout of the 6U8A. (*After* [1].)

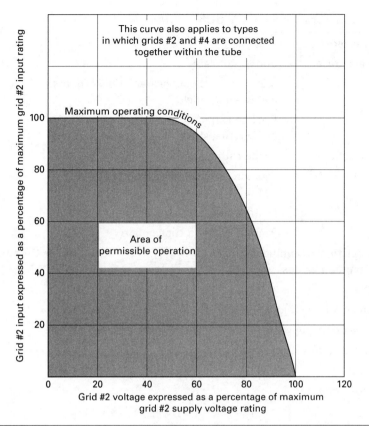

FIGURE 7.6 Grid #2 (screen grid) input rating curve. (*After* [1].)

TABLE 7.3 Key Parameters of the 6U8A. (*After* [1].)

General Characteristics		
Heater (AC/DC)		6.3 V, 0.45 A
Heater-cathode voltage	Peak value	±200 V maximum
	Average value	100 V maximum
Direct interelectrode capacitance (approx.), triode unit	Grid to plate	1.8 pF
	Grid to cathode, heater, and pentode elements	2.8 pF
	Plate to cathode, heater, and pentode elements	1.5 pF

(*continued*)

TABLE 7.3 Key Parameters of the 6U8A. (*After* [1].) (*Continued*)

General Characteristics

Direct interelectrode capacitance (approx.), pentode unit	Grid #1 to plate	0.01 pF maximum
	Grid #1 to cathode, heater, grid #2, grid #3, and internal shield	5 pF
	Plate to cathode, heater, grid #2, grid #3, and internal shield	2.6 pF
Direct interelectrode capacitance (approx.), other	Triode cathode to heater	3 pF
	Pentode cathode, pentode grid #2, and internal shield	3 pF
	Pentode grid #1 to triode plate	0.2 pF maximum
	Pentode plate to triode plate	0.1 pF maximum

Class A$_1$ Amplifier, Maximum Ratings (Design-Maximum Values)		Triode Unit	Pentode Unit
Plate voltage		330 V	330 V
Grid #2 (screen grid) voltage			See Figure 7.6
Grid #2 (screen grid) supply voltage			330 V
Grid #1 (control grid) voltage, positive-bias value		0 V	0 V
Plate dissipation		2.5 W	3 W
Grid #2 (screen grid) input			
	For grid # 2 voltages up to 165 V		0.55 W
	For grid #2 voltages between 165 and 330 V		See Figure 7.6

Characteristics	Triode Unit	Pentode Unit
Plate supply voltage	125 V	125 V
Grid #2 (screen grid) supply voltage		110 V
Grid #1 (control grid) voltage	–1 V	–1 V
Amplification factor	40	—
Plate resistance (approx.).	—	200 kΩ
Transconductance	7500 µmhos	5000 µmhos
Plate current	13.5 mA	9.5 mA
Grid #2 (screen grid) current		3.5 mA
Grid #1 (control grid) voltage (approx.) for plate current of 20 µA	–9 V	–8 V

the maximum rated screen grid supply voltage. From the 50 percent point to the full-rated value of supply voltage, the screen grid input must be decreased. The decrease in allowable screen grid input follows a parabolic curve. The rating chart is useful for applications utilizing either a fixed screen grid voltage or a series screen grid voltage-dropping resistor; specifically:

- When a fixed voltage is used, it is necessary only to determine that the screen grid input is within the boundary of the operating area on the chart at the selected value of screen grid voltage to be used.
- When a voltage-dropping resistor is used, the minimum value of resistor that will assure tube operation within the boundary of the curve can be determined by the relation

$$R_{g2} \geq \frac{E_{c2}(E_{cc2} - E_{c2})}{P_{c2}}$$

where:

R_{g2} = minimum value for the voltage-dropping resistor in ohms

E_{c2} = selected screen grid voltage in volts

E_{cc2} = screen grid supply voltage in volts

P_{c3} = screen grid input in watts corresponding to E_{c2}

6X4

The 6X4 is a miniature rectifier used in a variety of AC-operated circuits [1]. This tube, like other power-handling devices, should be adequately ventilated. The pinout of the seven-contact socket is shown in Figure 7.7. A generalized ratings chart for the 6X4 is shown in Figure 7.8. Selected tube parameters are listed in Table 7.4.

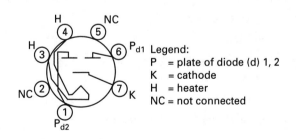

FIGURE 7.7   Terminal pinout of the 6X4. (*After* [1].)

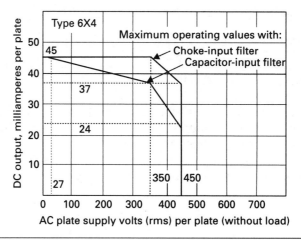

FIGURE 7.8 Ratings chart for the 6X4. (*After* [1].)

TABLE 7.4 Key Parameters of the 6X4. (*After* [1].)

General Characteristics		
Heater voltage	6.3 V	
Heater current	0.6 A	
Full-Wave Rectifier, Maximum Ratings (Design-Maximum Values)		
Peak inverse plate voltage	1250 V	
Peak plate current (per plate)	245 mA	
AC plate voltage supply (per plate, rms)	Figure 7.8	
Average output current (per plate)	Figure 7.8	
Typical Operation	**Choke-Input Filter**	**Capacitor-Input Filter**
AC plate supply voltage (each plate)	325 V	400 V
Filter input capacitor	10 μF	–
Effective plate supply impedance (each plate)	525 Ω	–
Filter input choke	–	10 H
Average output current	70 mA	70 mA
DC output voltage at input to filter (approx.)	310 V	340 V

5651A

The 5651A is a cold-cathode, glow-discharge, voltage-reference tube intended for use in DC power supplies. The device pinout for the seven-contact base is shown in Figure 7.9. Selected tube parameters are listed in Table 7.5.

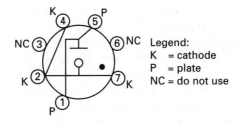

FIGURE 7.9 Terminal pinout of the 5651A. (*After* [1].)

A series resistor must always be used with the 5651A [1]. The resistance value must be chosen so that

- The maximum current rating of 3.5 mA is not exceeded at the highest anode supply voltage employed
- The minimum current rating of 1.5 mA is always exceeded when the anode supply voltage is at its lowest value

It is important to note that no connections to pins 3 and 6 should be made. Any potentials applied to these pins may cause erratic tube performance. The three pin terminals for the cathode (pins 2, 4, and 7) and the two pin terminals for the anode

TABLE 7.5 Key Parameters of the 5651A. (*After* [1].)

Maximum Ratings (Absolute Maximum Values)			
DC operating current (continuous), maximum			3.5 mA
DC operating current (continuous), minimum			1.5 mA
Ambient temperature range			−55 to 90°C
Characteristics and Operation Range Values		**Average**	**Maximum**
DC starting voltage		107 V	115 V[1]
DC operating voltage (variation from tube to tube)	At 1.5 mA	85 V	87 V
	At 2.5 mA	85.5 V	87.5 V
	At 3.5 mA	86.5 V	88.5 V
Regulation (1.5 mA to 3.5 mA)		–	3 V
Temperature coefficient of operating voltage (over ambient temperature range of −55 to 90°C)		−4 mV/°C	–
Percentage variation in operating voltage as a function of operating time; DC operating current = 2.5 mA; after initial 3-minute warm-up period		–	≥0.1 percent
Instantaneous voltage fluctuation (voltage jump)[2]		0.1 V	0.1 V
Circuit Values			
Shunt capacitor			0.02 μF

[1] A DC supply voltage of 115 V, minimum, should be provided to ensure "starting" throughout tube life.
[2] Defined as the maximum instantaneous voltage fluctuation at any current level within the operating current range.

(pins 1 and 5) offer several design possibilities for connection of the 5651A. Any pair of interconnected pins can be used as a jumper connection to a circuit common to either the cathode or to the anode. Such a jumper provides a means to open the circuit when the 5651A is removed from its socket. Under no circumstances should the current through any pair of interconnected pins exceed one ampere.

A warm-up period of three minutes should be allowed each time the equipment is turned on to ensure minimum voltage drift of the 5651A.

When a shunt capacitor is used with the 5651A, its value should be limited to 0.02 μF. A large value of capacitance may cause the tube to oscillate and thus give unstable performance.

Shielding should be utilized for the 5651A to ensure maximum stability when the tube is operated in the presence of strong magnetic fields.

5751

The 5751 is a high-μ twin triode used in high-gain amplifiers and industrial control devices. The device pinout for the nine-contact base is shown in Figure 7.10. Basic operating characteristics are charted in Figure 7.11. Selected tube parameters are listed in Table 7.6.

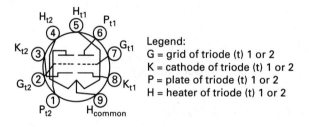

FIGURE 7.10   Terminal pinout of the 5751. (*After* [1].)

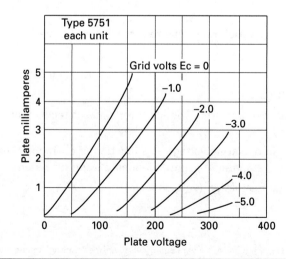

FIGURE 7.11   Characteristic curves of the 5751. (*After* [1].)

TABLE 7.6   Key Parameters of the 5751. (*After* [1].)

General Characteristics		
Heater (AC/DC)	Parallel connection	6.3 V, 0.35 A
	Series connection	12.6 V, 0.175 A
Heater-cathode voltage	Peak value	±100 V maximum
Class A$_1$ Amplifier, Each Unit, Maximum Ratings (Design-Maximum Values)		
Plate voltage		330 V
Plate dissipation		0.8 W
Grid voltage	Negative-bias value	55 V
	Positive-bias value	0 V
Bulb temperature (at hottest point on the envelope)		165°C
Characteristics		
Plate voltage	100 V	250 V
Grid voltage	–1 V	–3 V
Amplification factor	70	70
Plate resistance	58 kΩ	58 kΩ
Transconductance	1200 μmhos	1200 μmhos
Plate current	0.9 mA	1.0 mA

5879

The 5879 is a miniature type used as the input stage of a high-quality audio amplifier [1]. The nine-contact pinout is given in Figure 7.12. Device characteristic curves are shown in Figure 7.13. Selected tube parameters are listed in Table 7.7.

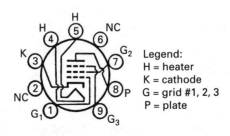

FIGURE 7.12   Terminal pinout of the 5879 tube. (*After* [1].)

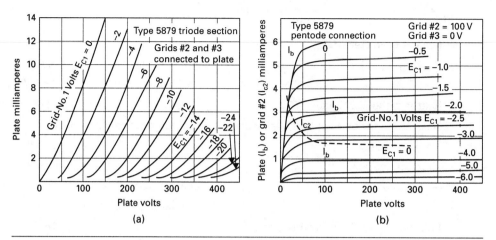

FIGURE 7.13 Characteristic curves for the 5879 tube: (*a*) triode section, (*b*) pentode section. (*After* [1].)

TABLE 7.7 Key Parameters of the 5879. (After [1].)

General Characteristics		
Heater (AC/DC)		6.3 V, 0.15 A
Peak heater-cathode voltage		±100 V maximum
Direct interelectrode capacitance (approx.)	Grid #1 to plate	0.11 pF maximum
	Grid #1 to cathode, heater, grid #2, and grid #3	2.7 pF
	Plate to cathode, heater, grid #2, and grid #3	2.4 pF
Class A₁ Amplifier, Maximum Ratings (Design-Maximum Values)		
Plate voltage		330 V
Plate dissipation		1.25 W
Grid #1 (control grid) voltage	Negative-bias value	55 V
	Positive-bias value	0 V
Grid #2 (screen grid) voltage		Note 1
Grid #2 (screen grid) supply voltage		330 V
Grid #2 (screen grid) input	For grid #2 voltages up to 165 V	0.25 W
	For grid #2 voltages between 165 V and 300 V	Note 1

(*continued*)

TABLE 7.7   Key Parameters of the 5879. (After [1].) (*Continued*)

Characteristics	
Plate voltage	250 V
Grid #3 (suppressor grid)	Connected to cathode at socket
Grid #2 (screen grid) voltage	100 V
Grid #1 (control grid) voltage	–3 V
Plate resistance (approx.)	2 mΩ
Transconductance	1000 µmhos
Plate current	1.8 mA
Grid #2 (screen grid) current	0.4 mA
Grid #1 (control grid) voltage (approx.) for plate current of 10 µA	–8 V
Maximum Circuit Value	
Grid #1 (control grid) circuit resistance	2.2 mΩ

Note:
1 See Figure 7.6 and the discussion of screen grid considerations outlined previously for the 6U8A tube.

6080

The 6080 is a low-µ, twin-power triode used as a regulator tube in DC power supplies. The device pinout for the octal base is shown in Figure 7.14. It is important that this tube, like other power-handling devices, be adequately ventilated. Basic operating characteristics are charted in Figure 7.15. Selected tube parameters are listed in Table 7.8.

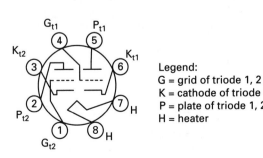

FIGURE 7.14   Terminal pinout of the 6080. (*After* [1].)

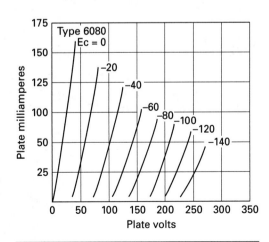

FIGURE 7.15   Characteristic curves of the 6080. (*After* [1].)

TABLE 7.8 Key Parameters of the 6080. (*After* [1].)

General Characteristics		
Heater voltage		6.3 V
Heater current		2.5 A
Heater-cathode voltage, peak		±300 V
Direct interelectrode capacitances (approx.)	Grid to plate (each unit)	8 pF
	Input (each unit)	6 pF
	Output (each unit)	2.2 pF
	Heater to cathode (each unit)	11 pF
	Grid of unit 1 to grid of unit 2	0.5 pF
	Plate of unit 1 to plate of unit 2	2 pF
DC Amplifier (Each Unit), Maximum Ratings (Absolute-Maximum Values)		
Plate voltage		250 V
Plate current		125 mA
Plate dissipation		13 W
Bulb temperature (at hottest point on the envelope surface)		200°C
Maximum Circuit Values		
Grid circuit resistance	For cathode bias operation	1 M Ω
	For fixed bias operation[1]	100 kΩ
	For combined fixed and cathode bias operation[2]	100 kΩ

[1] When fixed bias is used, the plate circuit should contain a protective resistance to provide a minimum drop of 15 V DC at the normal operating conditions.
[2] When combined fixed and cathode bias is used, the cathode bias portion should have a minimum value of 7.5 V DC at the normal operating conditions.

6973

The 6973 is a nine-pin, compact beam power tube used in high-fidelity audio equipment. The device pinout is given in Figure 7.16. Operating characteristics are charted in Figure 7.17. Selected tube parameters are listed in Table 7.9.

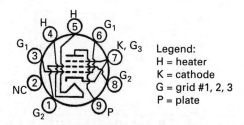

Legend:
H = heater
K = cathode
G = grid #1, 2, 3
P = plate

FIGURE 7.16 Terminal pinout of the 6973. (*After* [1].)

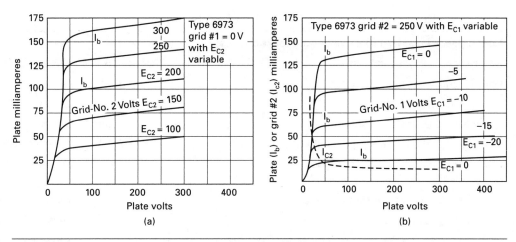

FIGURE 7.17 Characteristic curves of the 6973: (*a*) with the voltage on grid #2 variable, (*b*) with the voltage on grid #1 variable. (*After* [1].)

TABLE 7.9 Key Parameters of the 6973. (*After* [1].)

General Characteristics

Heater (AC/DC)		6.3 V, 0.45 A
Heater-cathode voltage	Peak value	±200 V maximum
	Average value	100 V maximum
Direct interelectrode capacitance (approx.), pentode unit	Grid to plate	0.4 pF maximum
	Grid #1 to cathode, heater, grid #2, and grid #3	9 pF
	Plate to cathode, heater, grid #2, and grid #3	6 pF

Push-Pull Class AB$_1$ Amplifier, Maximum Ratings (Design-Maximum Values)

Plate voltage	440 V
Grid #2 (screen grid) voltage	330 V
Plate dissipation	12 W
Grid #2 (screen grid) input	2 W
Envelope temperate at hottest point	250°C

Typical Operation, Fixed Bias (values are for two tubes)

Plate supply voltage	250 V	350 V	400 V
Grid #2 (screen grid) supply voltage	250 V	280 V	290 V

(continued)

TABLE 7.9   Key Parameters of the 6973. (*After* [1].) (*Continued*)

Typical Operation, Fixed Bias (values are for two tubes)

Grid #1 (control grid) voltage	−15 V	−22 V	−25 V
Peak AF grid #1 to grid #1 voltage	30 V	44 V	50 V
Zero signal plate current	92 mA	58 mA	50 mA
Maximum signal plate current	105 mA	106 mA	107 mA
Zero signal grid #2 (screen grid) current	7 mA	3.5 mA	2.5 mA
Maximum signal grid #2 (screen grid) current	16 mA	14 mA	13.7 mA
Effective load resistance (plate to plate)	8000 Ω	7500 Ω	8000 Ω
Maximum signal output	12.5 W	20 W	24 W
Grid #1 (control grid) circuit resistance for fixed bias operation		500 kΩ	

Push-Pull Class AB$_1$ Amplifier With Grid No. 2 of Each Tube Connected to a Tap on the Output Transformer

(Design Maximum Values)

Plate and grid #2 supply voltage	410 V
Plate dissipation	12 W
Grid #2 input	1.75 W
Envelope temperature at hottest point	250°C

Typical Operation (values are for two tubes)	**Fixed Bias**	**Cathode Bias**
Plate supply voltage	375 V	370 V
Grid #2 (screen grid) supply voltage	Note 1	Note 1
Grid #1 (control grid) voltage	−33.5	−
Cathode bias resistor	−	355 Ω
Peak AF grid #1 to grid #1 voltage	67 V	62 V
Zero signal cathode current	62 mA	74 mA
Maximum signal cathode current	95 mA	84 mA
Effective load resistance (plate to plate)	12.5 kΩ	13 kΩ
Maximum signal output	18.5 W	15 W
Grid #1 (control grid) circuit resistance for fixed bias operation	100 kΩ	

Note:

1 Obtained from taps on the primary winding of the output transformer. The taps are located on each side of the center tap (B+) so as to apply 50% of the plate signal voltage to grid #2 of each output tube.

7025 (12AX7)

The 7025 is a miniature type used as a phase inverter or resistance-coupled amplifier in high-fidelity audio amplifiers [1]. This type is identical with the 12AX7A (ECC83) tube, except that it has a controlled equivalent noise and hum characteristic. Each triode is independent of the other except for the common heater. The nine-contact pinout is given in Figure 7.18. Device characteristic curves are shown in Figure 7.19. Selected tube parameters are listed in Table 7.10.

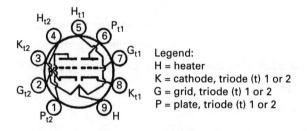

FIGURE 7.18   Terminal pinout of the 7025 tube. (*After* [1].)

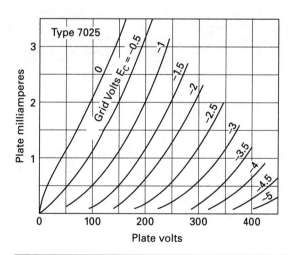

FIGURE 7.19   Characteristic curves for the 7025 tube. (*After* [1].)

TABLE 7.10 Key Parameters of the 7025. (*After* [1].)

General Characteristics			
Heater (AC/DC)	Parallel connection		6.3 V, 0.30 A
	Series connection		12.6 V, 0.15 A
Heater-cathode voltage	Peak value		±200 V maximum
	Average value		100 V maximum
Direct interelectrode capacitance (approx.)		Unit #1	Unit #2
	Grid to plate	1.7 pF	1.7 pF
	Grid to cathode and heater	1.6 pF	1.6 pF
	Plate to cathode and heater	0.46 pF	0.34 pF
Class A$_1$ Amplifier, Each Unit, Maximum Ratings (Design-Maximum Values)			
Plate voltage			330 V
Plate dissipation			1.2 W
Grid voltage	Negative bias value		55 V
	Positive bias value		0 V
Equivalent Noise and Hum Voltage Referenced to Each Grid (each unit)			
Average value (rms)[1]			1.8 µV
Maximum value (rms)[2]			7 µV

[1] Measured in "true rms" units under the following conditions: heater volts (AC) = 6.3 V (parallel connection); center tap of heater transformer connected to ground; plate supply = 250 V; plate load resistor = 2700 Ω; cathode bypass capacitor = 100 µF; grid resistor = 0 Ω; and amplifier covering frequency range 25 Hz to 10 kHz.
[2] Same conditions as for "average value," except the cathode resistor is unbypassed and the grid resistor = 50 kΩ.

7199

The 7199 is a miniature type used in high-quality, high-fidelity audio equipment, particularly in phase splitters, tone-control amplifiers, and high-gain voltage amplifiers. The nine-contact pinout is given in Figure 7.20. Characteristic curves for the device are illustrated in Figure 7.21. In direct-coupled, phase-splitter circuits, the pentode unit should drive the triode unit. Selected tube parameters are listed in Table 7.11.

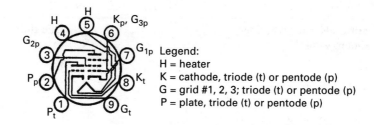

FIGURE 7.20 Terminal pinout of the 7199. (*After* [1].)

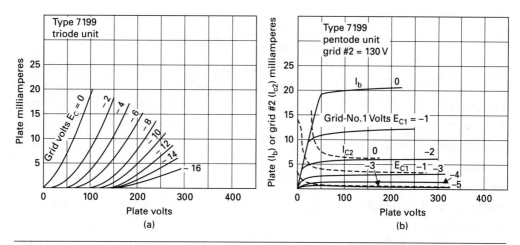

FIGURE 7.21 Characteristic curves of the 7199: (a) triode unit, (b) pentode unit. (*After* [1].)

TABLE 7.11 Key Parameters of the 7199. (*After* [1].)

General Characteristics

Heater (AC/DC)		6.3 V, 0.45 A
Heater-cathode voltage	Peak value	±200 V maximum
	Average value	100 V maximum
Direct interelectrode capacitance (approx.), triode unit	Grid to plate	2 pF
	Grid to cathode and heater	2.3 pF
	Plate to cathode and heater	0.3 pF
Direct interelectrode capacitance (approx.), pentode unit	Grid #1 to plate	0.06 pF maximum
	Grid #1 to cathode, heater, grid #2, grid #3, and internal shield	5 pF
	Plate to cathode, heater, grid #2, grid #3, and internal shield	2 pF

(*continued*)

TABLE 7.11   Key Parameters of the 7199. (*After* [1].) (*Continued*)

Class A₁ Amplifier, Maximum Ratings (Design-Maximum Values)		Triode Unit	Pentode Unit
Plate voltage		330 V	330 V
Grid #2 (screen grid) voltage			See Figure 7.6 and related text
Grid #2 (screen grid) supply voltage			330 V
Grid #1 (control grid) voltage, positive-bias value		0 V	0 V
Plate dissipation		2.4 W	3 W
Grid #2 (screen grid) input			
	For grid #2 voltages up to 165 V		0.6 W
	For grid #2 voltages between 165 and 330 V		See Figure 7.6 and related text
Characteristics		**Triode Unit**	**Pentode Unit**
Plate supply voltage		215 V	220 V
Grid #2 (screen grid) supply voltage			130 V
Grid #1 (control grid) voltage		–8.5 V	—
Amplification factor		17	—
Plate resistance (approx.).		8.1 kΩ	400 kΩ
Transconductance		2100 μmhos	7000 μmhos
Plate current		9 mA	12.5 mA
Grid #2 (screen grid) current			3.5 mA
Grid #1 (control grid) voltage (approx.) for plate current of 10 μA		–40 V	—

7868

The 7868 is a nine-pin, compact beam power tube used in high-fidelity audio equipment. The pinout of the nine-contact novar socket device is given in Figure 7.22. This tube, like other power-handling devices, should be adequately ventilated. Operating characteristics are charted in Figure 7.23. Selected tube parameters are listed in Table 7.12.

Significant performance benefits can be achieved by using the output transformer screen tap configuration in a push-pull audio stage. The dramatic improvement in distortion will be documented in later chapters. This improvement, however, does not come without a cost. As shown in Table 7.12 the decrease in output power between the fixed screen supply and the transformer tap is easily 25 percent. As detailed in Chapter 11, this tradeoff is usually acceptable. As mentioned before, circuit design is an exercise in tradeoffs.

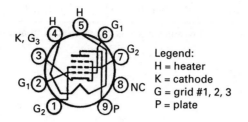

FIGURE 7.22 Terminal pinout of the 7868. (*After* [1].)

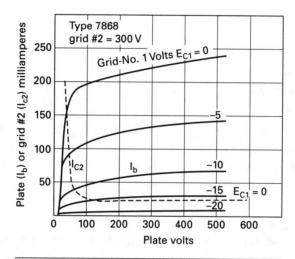

FIGURE 7.23 Characteristic curves of the 7868. (*From* [1].)

TABLE 7.12 Key Parameters of the 7868. (*After* [1].)

General Characteristics

Heater (AC/DC)		6.3 V, 0.8 A
Heater-cathode voltage	Peak value	±200 V maximum
	Average value	100 V maximum
Direct interelectrode capacitance (approx.), pentode unit	Grid to plate	0.15 pF maximum
	Grid #1 to cathode, heater, grid #2, and grid #3	11 pF
	Plate to cathode, heater, grid #2, and grid #3	4.4 pF

(*continued*)

TABLE 7.12 Key Parameters of the 7868. (*After* [1].) (*Continued*)

Push-Pull Class AB$_1$ Amplifier, Maximum Ratings (Design-Maximum Values)

Plate voltage	550 V
Grid #2 (screen grid) voltage	440 V
Average cathode current	90 mA
Plate dissipation	19 W
Grid #2 (screen grid) input	3.3 W (Note 1)
Envelope temperature at hottest point	240°C

Typical Operation, Fixed Bias (values are for two tubes)

Plate supply voltage	350 V	400 V	450 V
Grid #2 (screen grid) supply voltage	350 V	350 V	350 V
Grid #1 (control grid) voltage	−15.5 V	−16 V	−16.5 V
Peak AF grid #1 to grid #1 voltage	31 V	32 V	33 V
Zero signal plate current	72 mA	64 mA	60 mA
Maximum signal plate current	130 mA	135 mA	142 mA
Zero signal grid #2 (screen grid) current	9.5 mA	8 mA	7.2 mA
Maximum signal grid #2 (screen grid) current	32 mA	28 mA	26 mA
Effective load resistance (plate to plate)	6600 Ω	6600 Ω	6600 Ω
Maximum signal output	30 W	34 W	38 W

Push-Pull Class AB$_1$ Amplifier
Grid #2 of Each Tube Connected to Tap on Plate Winding of Output Transformer

Typical Operation (values are for two tubes)	Fixed Bias	Cathode Bias
Plate supply voltage	400 V	425 V
Grid #2 supply voltage	Note 2	Note 2
Grid #1 voltage	−20 V	−
Cathode bias resistor	−	185 Ω
Peak AF grid #1 to grid #1 voltage	41 V	42 V
Zero signal plate current	60 mA	88 mA
Maximum signal plate current	115 mA	100 mA
Zero signal grid #2 current	8 mA	12 mA
Maximum signal grid #2 current	18 mA	16 mA
Effective load resistance (plate-to-plate)	6600 Ω	6600 Ω
Total harmonic distortion	2.5%	3.5%
Maximum signal power output	23 W	21 W

Notes:
1 Grid #2 (screen grid) input may reach 6 W during peak input levels.
2 The grid #2 supply voltage is obtained from taps on the primary winding of the output transformer. The taps are located on each side of the center tap (B+) so as to apply 50% of the plate signal voltage of each output tube.

Application Considerations

The data given in the charts, drawings, and tables in this chapter provide a starting point for circuit design. A number of variations in circuit configurations are possible for any given application, and sometimes a circuit designed strictly "by the book" fails to perform as expected. As circuit design is often an iterative process, the data given in this chapter should be carefully considered during the design stage, and then again after the circuit has been optimized.

Variations in operating parameters among tubes of a given type number are typically small for NOS devices. In the case of tubes still being manufactured, certain changes in operating parameters relative to NOS devices may be found. Consult the device manufacturer to be sure.

It is worth noting that tubes used in the push-pull output stage of an audio amplifier will need to be balanced for best performance. The methods of achieving balance fall into two general categories: 1) utilize a mechanism to adjust the relative drive of each stage to achieve balance, or 2) employ tubes whose performance has been matched to provide essentially identical output. Both approaches have their merits. Generally speaking, however, the author prefers the matched tube method since it simplifies circuit design and setup. Experience has shown that a matched pair does not deviate appreciably over its useful operating life. Near the end of useful life, the tubes can exhibit significant performance differences. Of course, by then it is time to replace the devices. Some tube types are difficult to find commercially available. It may be worthwhile, therefore, to design circuits around devices that are still being manufactured, or ones that have multiple sources of NOS stock.

Apart from tubes, other devices that can be hard to locate with the needed specifications for a given circuit include power transformers and audio output transformers. As with NOS tubes, the reader may want to consider component availability a key design item.

Chapter 8

Project 1: Power Supply

Project 1 is a power supply for any number of applications, including providing supply voltages for the projects described in the following chapters. The design objectives are as follows:

- Fixed output B+ supply in the range of 180 V to 410 VDC at up to 200 mA
- Regulated supply option
- Filament voltages for various heater requirements
- Total project cost for parts below $500

The basic circuit options for the power supply are shown in Figure 8.1. The regulator circuit is shown in Figure 8.2.

Circuit Description

A power supply does not produce power for its load. Rather, it takes power from a source and conditions the power so that the load can operate properly and without damage. This conditioning includes removing ripple components and noise, and in the case of a regulated power supply, output voltage variations despite changes in the input line voltage and/or changes in the load current.

The power supply circuits shown in Figure 8.1 use either a 5BC3 or 6X4 tube as the rectifier [1]. The 5BC3 is a directly heated novar device intended for relatively high DC requirements, whereas the 6X4 is intended for more moderate DC requirements. On the 6X4, the cathode and heater structures are separated, which simplifies heater requirements in certain applications.

In each rectifier circuit, the 120 V AC input power is applied to the primary of step-up power transformer T1. The two plate sections of the rectifier tube are connected to opposite ends of the center-tapped secondary winding of transformer T1. With respect to the grounded center tap, the voltage applied to each plate of the rectifier tube, therefore, is 180° out of phase with that applied to the other plate. With an external load connected to the rectifier cathode, pulses of current flow alternately

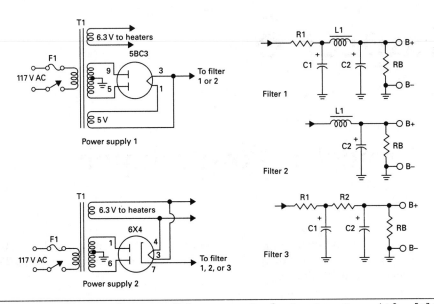

FIGURE 8.1   Schematic diagram of the power supply circuit options. (*After* [1].)

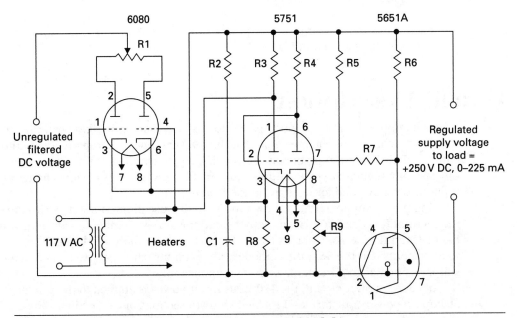

FIGURE 8.2   Power supply regulation circuit. (*After* [1].)

to one plate and then to the other plate for each half-cycle of the AC input power. This 120 Hz pulsating current develops a positive DC voltage across the load.

Removal of virtually all the 120 Hz ripple component from the DC output can be accomplished through a connection to a suitable filter network between the rectifier output (cathode) and the load. Either Filter 1 or Filter 2 in Figure 8.1 provides adequate filtering of the 5BC3 circuit. Any one of the three filter networks is satisfactory for use with the 6X4 circuit. Filter 3 is not recommended for use with the 5BC3 because the use of the two resistors (R1 and R2) in series with the relatively high-output voltage may result in excessive power loss.

As shown in Table 8.1, a wide range of DC output voltages can be obtained for various values of load current. Design criteria include the following:

- Selection of power transformer T1
- Type of filter network
- Values of filter choke L1
- Value of resistors R1 and R2, if used

Note that the voltage and current values shown in Table 8.1 are based on calculations and supporting measurements as documented in [1]. These measurements were not repeated by the author.

Bleeder resistor RB may be omitted if an external load is permanently connected across the output terminals. Bleeder current should be approximately 10 percent of the load current. Note that if a bleeder resistor is not used and the power supply is energized with the load disconnected, dangerous voltages will be present in the power supply for some time after primary power has been removed. For this reason, it is recommended that a bleeder resistor be used in all power supply designs.

Figure 8.2 shows a series regulator circuit using a 5651A as the voltage reference tube, a type 6080 as the series regulator tube, and a type 5751 as the control tube. In this circuit, the 5651A supplies a fixed reference voltage between the grid of the first unit of the 5751 and its cathode return. Changes in supply voltage to the load are amplified by the 5751, which is connected as a two-stage DC amplifier to control the voltage drop through the 6080 tube. The resulting output voltage is essentially independent of changes in load current.

The voltage regulation of this supply, operated at a fixed line voltage of 120 V and an output voltage of 250 V, is approximately 0.5 V over the current range of 0 to 225 mA. At full current, the regulation for a variation of ± 10 percent in line voltage is typically less than 0.5 V.

The input voltage to the regulator circuit (with the component values shown) can range from approximately 325 V to 400 V or so at 225 mA load current.

The socket connections shown for the 5651A voltage reference tube are made so that removal of the 5651A from its socket will open (disconnect) the load. This is important to prevent a runaway condition whereby the output voltage of the regulator is run up to the maximum value of the unregulated input. Note also that pins 3 and 6 of the 5615A should not be used.

TABLE 8.1 Power Supply Circuit Options (*After* [1].)

Rectifier Type	Transformer Secondary	Choke	R1	R2	C1, C2	Filter Type	Voltage, V	Current, mA
5BC3	300-0-300 V	7 H, 140 mA, 165 Ω	33 Ω, 5W		40 µF, 450 V	1	360	60
							340	80
							320	120
						2	235	60
							230	80
							215	120
	400-0-400 V	4 H, 200 mA, 145 Ω	56 Ω, 10 W		40 µF, 600 V	1	450	120
							425	160
							410	200
						2	310	120
							300	160
							280	200
6X4	300-0-300 V	12 H, 80 mA, 375 Ω	500 Ω, 5W	500 Ω, 3W	40 µF, 450 V	1	350	20
							300	40
							260	60
						2	250	20
							230	40
							220	60
						3	345	20
							300	40
							250	60
	240-0-240	12 H, 80 mA, 375 Ω	500 Ω, 5W	500 Ω, 3W	40 µF, 450 V	1	265	20
							225	40
							190	60
						2	200	20
							180	40
							170	60
						3	260	20
							220	40
							180	60

Parts List

Table 8.2 lists the electronic components needed to build a power supply with the following specifications:

- Unregulated output supply voltage = 400 V, 200 mA, using Filter 2 as shown in Figure 8.1
- Regulated output voltage = 250 V to 275 V DC at 0 to 200 mA
- 6.3 V AC filament supply

TABLE 8.2   **Parts List for the Initial Build of the High-Voltage Power Supply**

Part	Description	Quantity	Manufacturer	Part No.	Allied Stock No.
C1, C2[1]	100 μF, 450 V, electrolytic[2]	2	Illinois Capacitor	22X50MM	613-0836
C2[1]	0.1 μF, 600 V	1	Vishay	715P10456LD3	507-0328
RB[1]	100 kΩ, 1 W[2]	2	Ohmite	OM1045E	296-4793
R1	50 Ω, 25 W rheostat	1	Ohmite	RHS50RE	296-4113
R2, R5	12 kΩ, 2 W	2	Vishay	PR02000201202JR500	648-0160
R3, R4	470 kΩ, 0.5 W	2	Ohmite	OL4745E	296-4789
R6	68 kΩ, 2 W	1	Ohmite	OY683KE	296-2384
R7	1 mΩ, 0.5 W	1	Ohmite	OL1055E	296-4971
R8	15 kΩ, 2 W	1	Ohmite	OY153KE	296-2333
R9	10 kΩ, potentiometer	1	Clarostat	RV4NAYSD103A	753-1310
L1	Choke: 5 H 300 mA 55 Ω	1	Hammond[3]	193LP	Newark
T1	Power transformer: Primary– 120 V Secondary– 400-0-400 V rms, 288 mA 6.3 V, 5 A 5 V, 3 A 5 V, 1.2 A	1	Hammond[3]	300BX	Newark

(continued)

TABLE 8.2 Parts List for the Initial Build of the High-Voltage Power Supply (*Continued*)

Part	Description	Quantity	Manufacturer	Part No.	Allied Stock No.
V1	5BC3 tube	1			The Tube Store[4]
V2	6080 tube	1			Tube Depot[4]
V3	5751 tube	1	Sovtek, JAN Philips, others		Tube Depot
V4	5651A tube	1			Tube Store

[1] Devices for the unregulated power supply filter circuit.
[2] Electrolytic capacitors are not commonly available at voltage ratings of 600 V or more. As a substitute, two 100 μF, 450 V capacitors are used in series, with 100 kΩ, 1 W resistors in parallel across each. Part numbers for this approach are given.
[3] Hammond Manufacturing Company, Inc., 475 Cayuga Rd., Cheektowaga, NY,14225,USA: www.hammondmfg.com.
[4] Vacuum tubes are available from a number of suppliers, including Tube Depot (www.tubedepot.com), The Tube Store (http://thetubestore.com), Tube World (www.tubeworld.com), and Antique Electronic Supply (www.tubesandmore.com).

The filter capacitor (C2) in Figure 8.1 is implemented in this application as two capacitors in series, with parallel resistors for voltage distribution and to ensure capacitor discharge when de-energized. With a voltage in excess of 400 V DC on C1, a safety margin of at least 100 V is recommended for the filter capacitor. It was found to be economical to use two 100 μF 450 V electrolytic capacitors in series rather than a single 50 μF 600 V electrolytic. The resistors (100 KΩ 1 W) are placed in parallel with each capacitor in order to equalize voltage distribution across the devices.

A component list is provided, along with manufacturer part numbers and stock numbers from one of the major parts houses.[1] These parts are, of course, available from a number of manufacturers and suppliers. The details given here are for the convenience of the reader. Considerations relating to chassis components and assembly techniques are addressed in other chapters.

At the time of this writing, the 5751 tube was being manufactured by one or more tube companies. It appeared that the 5BC3, 6080, and 5651A were available only from new old stock (NOS). Also, the 6X4 appeared to be available only from NOS.

Regarding rheostat R1, the device specified in Table 8.2 is overrated insofar as wattage is concerned. The Ohmite RHS50RE is rated for 25 W, but for this circuit, a wattage rating of half that would suffice. The device chosen is specified, rather, for its rated operating voltage. It can be seen from Figure 8.2 that R1 carries the full input voltage, which can range as high as 400 V to ground. As such, the rated operating voltage must be considered.

[1] Allied Electronics, 7151 Jack Newell Blvd. S., Fort Worth, Texas, 76118, USA, https://www.alliedelec.com.

Construction Considerations

The power supply described in this chapter can be constructed in any number of ways. For the purposes of this example, this unit was hand-wired using terminal strips on an aluminum chassis. A parts list for the chassis components is given in Table 8.3. Bulk supplies needed for project assembly are listed in Chapter 6 (Table 6.6).

Construction of the power supply utilizes conventional hand-wiring techniques. Details are shown in Figure 8.3.

The chassis used by the author in this implementation is oversized (17-inch by 14-inch by 3-inches deep) and designed to accommodate the project described in this chapter and the power amplifiers described in Chapters 10 and 11. As such, this is a test bed, intended to minimize repetitive sheet metal work and economize on components, in particular, the power supply transformers. Note that the chassis and related components listed in Table 8.3 are intended for just the power supply described in this chapter. Construction techniques are essentially identical for either the single-purpose chassis or the test bed.

One of the benefits of the test bed approach is that it affords the opportunity to easily make circuit modifications and compare the performance of different power amplifier and power supply designs.

Documentation detailing the transformer output lead color codes are provided with the device, or they can be downloaded from the manufacturer's website. For the power transformer listed in Table 8.2, a variety of primary input voltages can be accommodated, from a low of 100 V AC to a high of 240 V AC. It is likely that some

TABLE 8.3   Chassis Parts List for the Initial Build Circuit

Description	Quantity	Manufacturer	Part No.	Allied Stock No.	Notes
Chassis, 9.5-in × 5-in × 2.5-in[1]	1	Bud	CU-592-A	736-3611	
9-pin tube socket	1	Belton	SK-B-VT9-ST-2		Tube Depot
9-pin novar socket	1	Belton	SK-9PINX		Tube Depot
7-pin tube socket	1	Belton	SK-7PIN		Tube Depot
Octal tube socket	1	Belton	SK-B-VT8-ST		Tube Depot
AC line fuse holder	1	Bussmann	HTB-32I-R	740-0652	
Fuse, 3 A, slo-blow	1	Bussmann	MDL-3-R	740-0748	
AC power cord	1	Alpha Wire	615 BK078	663-7020	6 ft. power cord
Power switch	1	Eaton	7504K4	757-4201	
Chassis feet	4	Bud	F-7264-A	736-7264	
Terminal strip	8	Cinch	54A	750-6628	

[1] This chassis is sufficient for the regulator circuit. A considerably larger chassis is needed to mount the power transformer and choke; e.g., Hammond 1441-29BK3 (12-inch by 10-inch, 2 inches deep).

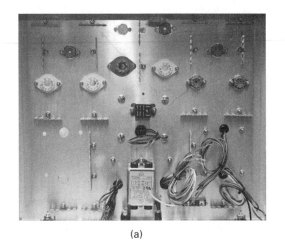

(a)

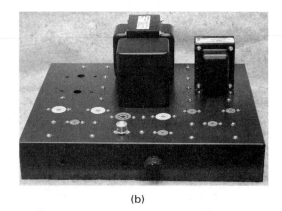

(b)

(c)

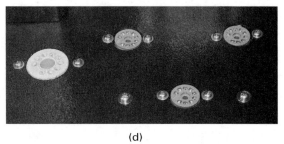

(d)

FIGURE 8.3 The project test bed: (*a*) bare chassis bottom view; (*b*) bare chassis top view; (*c*) power transformer, choke, and output transformer; (*d*) tube socket mounting detail.

number of transformer leads will not be used (in particular, the filament center-tap leads). Be sure to properly terminate the wires for the unused leads, either by connecting them to an empty terminal strip or by applying a wire-nut to each lead, as shown in Figure 8.4.

It is important to note that the 5BC3A rectifier tube heater draws a rated current of 3 A. For that reason, the proper 5 V secondary output winding on the power transformer must be used. There are two available 5 V output windings on the specified Hammond transformer, one rated for 1.2 A and the other rated for 3 A.

The completed unit is documented in Figure 8.5.

In the process of constructing and then testing the regulated power supply, the author explored and implemented the changes detailed in the following sections.

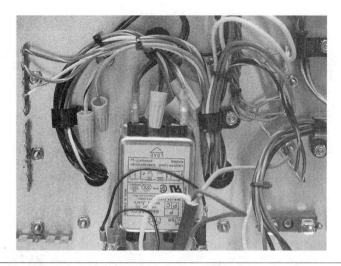

FIGURE 8.4   Termination of unused transformer leads

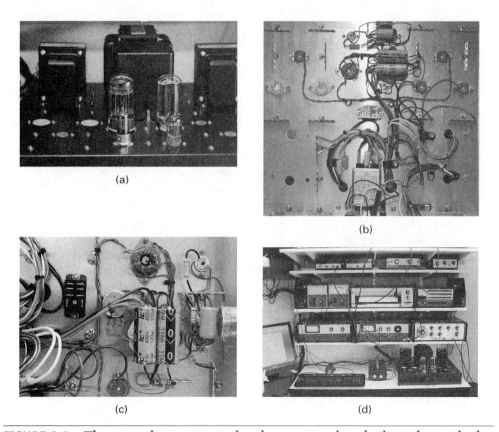

FIGURE 8.5   The general-purpose regulated power supply as built on the test bed chassis: (*a*) top view, (*b*) bottom view, (*c*) detail view, (*d*) power supply on the test bench powering the preamplifier described in Chapter 9.

DC Return Path Issue

The 5651A voltage reference tube is wired into the circuit so that should the tube be removed from its socket, the output of the power supply will be disconnected from the load. This is a protective mechanism, as without the voltage reference, the output of the supply would increase to the maximum level available from the filtered DC input (less any voltage drop across the 6080 series regulator device). The approach shown in Figure 8.2 requires that the common leg of the supply return through the 5651A socket protection circuit. This was found to be suboptimal for a variety of reasons.

To achieve a measure of protection from failure or removal of the 5651A, a zener diode was added to the circuit to serve as a backup for the tube. The 1 MΩ resistor R7 was replaced with two 470 KΩ resistors in series, and a zener diode was placed between the junction of the resistors and ground. A zener rated for 98 V[2] was used to keep the zener well outside the typical operating point of the 5651A but still provide protection in the event of removal of the reference tube or failure of the tube. This change permits the DC supply and regulator to be referenced to ground at all times while still protecting the load from a runaway regulator condition.

R9 Voltage Adjust Issue

The voltage adjust potentiometer (R9) allows for a wide range of output voltages. The measured minimum output was 250 V DC, and the measured maximum output voltage was approximately 350 V DC with no load connected and an input supply voltage of 400 V DC. The maximum voltage output is achieved with R9 at minimum resistance, which in effect grounds pin 8 of the 5751 tube. There are three problems with this:

- With pin 8 essentially grounded, the 350 V DC output is placed across R5 (12 kΩ, 2W), which causes the resistor to overheat.
- With pin 8 at ground potential, the possibility exists for exceeding the maximum rated heater-to-cathode voltage on one element of the 5751.
- The adjustment range (250 V to 350 V, no load) exceeds the practical range needed for this supply.

The modification to address these issues involves replacing the 10 kΩ potentiometer with a lower value and padding out the extremes to prevent misadjustment. Specifically, a 4.7 kΩ 2 W resistor is connected from pin 8 of the 5751 to the high end of R9, which has been changed in value to 5 kΩ. A 330 Ω 2 W resistor is connected from the low end of R9 to ground.[3] The total resistance from pin 8 to ground is the same as before with the

---

[2] A 98 V zener rating was achieved by using two devices in series: one rated for 51 V at 5 W (ON Semiconductor 1N5369BG, Allied stock # 568-0593) and the other rated for 47 V at 5 W (ON Semiconductor 1N5368BG, Allied stock # 568-0592).

[3] Part numbers are as follows: 5 kΩ 2 W potentiometer, Honeywell/Clarostat RV4NAYSD502A, Allied stock # 753-1445; 4.7 kΩ 2 W resistor, Ohmite OY472KE, Allied stock # 296-2369; 330 Ω 2 W resistor, Ohmite OY331KE, Allied stock # 296-2353.

R9 wiper on the low end (10 kΩ, approximately). However, the resistance from pin 8 to ground is at least 5 kΩ when the R9 wiper is on the high end. This change prevents the supply from running open at any setting of R9. It also improves the voltage adjustment resolution of the potentiometer.

And speaking of pin 8 of the 5751, note that pin 8 is tied to one side of the 6.3 V AC filament supply through connection to pins 4 and 5. This is necessary to avoid exceeding the maximum filament-to-cathode voltage rating of the 5751, which is ±100 V; for the 6080 regulator tube, the rating is ±300 V. By connecting pin 8 to pin 4/5 on the 5751, the voltage differential is maintained within a safe range.

At the nominal output of 250 V DC, about half the output voltage potential (125 V) can appear at either or both pins 3 and 8 of the 5751 due to the voltage divider network of R2/R8 for pin 3 and R5/R9 for pin 8.

Going back to the R9 voltage adjustment situation described previously, without the modification described in this section, it would be possible to exceed the maximum rated heater-to-cathode voltage through adjustment of R9. The resistive divider described in this section prevents this condition.

It is important to note that because of the DC potential on the filament, a separate 6.3 V AC winding must be used for any other heater circuits that may be powered by this supply. If a separate 6.3 V AC filament winding is not available from the transformer (as is the case with the Hammond 300BX used in the test bed), a separate filament transformer is required. The transformer would need to handle the rated load of the 6080 and 5751 heaters, which is approximately 3 A. A device such as the Hammond 167Q6 should suffice.

Initial Checkout

After the power supply has been assembled, the following steps are recommended to confirm proper construction. These values assume no load connected to the power supply.

1. Remove AC power from the unit (unplug the AC line cord).
2. Remove all tubes.
3. Using an ohmmeter, confirm that the resistance values listed in Table 8.4 are observed on the tube socket connections. All readings are with reference to ground. Readings within ±5 percent are considered normal.
4. If any deviations from the expected values listed in Table 8.4 are observed, recheck wiring and components for correct installation. Do not advance to the next step until the resistance measurements are within tolerance.
5. Because of the nature of the power supply design, no meaningful voltage measurements can be made on the circuit with the tubes removed. It is intuitive that all measurements (except for the filaments and rectifier plates) would be zero.
6. Install all tubes.
7. Set rheostat R1 to its center position.

TABLE 8.4 Typical Ohmmeter Readings with Power Removed and Tubes Removed

Tube	Pin No.	Typical Reading	Notes
5BC3A, rectifier	1	~200 kΩ	Varies due to filter capacitors in circuit (heater/cathode)
	2	–	
	3	~200 kΩ	Varies due to filter capacitors in circuit (heater/cathode)
	4	–	
	5	~46 Ω	Plate 1
	6	–	
	7	–	
	8	–	
	9	~46 Ω	Plate 2
6080 series regulator	1	480 kΩ	
	2	~200 kΩ	
	3	10.2 kΩ to 12 kΩ	Varies depending on setting of output voltage control R9[1]
	4	480 k	
	5	~200 kΩ	Varies due to filter capacitors in circuit
	6	10.2 kΩ to 12 kΩ	Varies depending on setting of output voltage control R9[1]
	7	~4 kΩ to 7.6 kΩ	Varies depending on setting of output voltage control R9[1] (heater)
	8	~4 kΩ to 7.6 kΩ	Varies depending on setting of output voltage control R9[1] (heater)
5751 error amplifier	1	480 kΩ	
	2	480 kΩ	
	3	~10 kΩ	Varies a small amount with the setting of R9
	4	~4 kΩ to 7.6 kΩ	Varies depending on setting of output voltage control R9[1] (heater)
	5	~4 kΩ to 7.6 kΩ	Varies depending on setting of output voltage control R9[1] (heater)
	6	480 kΩ	
	7	1 mΩ	
	8	~4 kΩ to 7.6 kΩ	Varies depending on setting of output voltage control R9[1]
	9	~4 kΩ to 7.6 kΩ	Varies depending on setting of output voltage control R9[1] (heater)

(continued)

TABLE 8.4   Typical Ohmmeter Readings with Power Removed and Tubes Removed (*Continued*)

Tube	Pin No.	Typical Reading	Notes
5651A reference tube	1	–	
	2	0 Ω	
	3	–	
	4	–	
	5	79 kΩ	
	6	–	
	7	–	

[1] These readings apply to the modified R9 voltage divider described previously.

8. Set voltage control potentiometer R9 to its center position.
9. Remove any load from the regulator.
10. Connect the AC line cord and switch the power on. Observe the tubes for any signs of overheating or other unexpected behavior. The filaments should be lit. Remove power if any problems are detected.
11. Carefully measure the voltage at the input to the regulator. It should be approximately 400 V DC.
12. Carefully measure the voltage at the output of the regulator. It should be between 250 V and 300 V DC.
13. Adjust potentiometer R9 and confirm that the output voltage varies with the control. Important: if you are using the basic circuit without the modifications described previously to prevent possible misadjustment of R9, keep the potentiometer within the center of its range.
14. Rheostat R1 is used to balance the two sections of the 6080 series regulator tube. Balance is indicated when the measured voltages at each end of the rheostat (i.e., each plate of the 6080) are identical when working into the characteristic load. This adjustment must be made with extreme care, as potentially lethal voltages are present. After adjustment, R1 should remain close to center position. If significant deviation from center is required to achieve balance, check the circuit and/or consider trying a replacement tube.
15. After these preliminary tests have been successfully completed, measurements should be taken on the power supply as detailed in the next section.

Measured Performance

Upon completion of the power supply, key measurements were taken, as documented in Table 8.5.

After confirming these measurements, turn off the power supply and remove the AC line cord. Using a voltmeter, check the wiper arm of R1 to confirm that the voltage is zero. The power supply should fully discharge within about five seconds.

TABLE 8.5 Measured Performance of the General-Purpose Regulated Power Supply with No Load

Conditions	Parameter	Measured Value	Notes
AC line voltage		118 VAC	
Voltage measurements under conditions specified	Regulated supply output	250 VDC	
	6080 pin 2 (plate)	407 VDC	
	6080 pin 5 (plate)	407 VDC	
	6080 pin 1/4 (grids)	136 VDC	
	5651A pin 5 (plate)	83 VDC	
Hot test of reference tube failure	Output with 5651A removed from socket	275 VDC	Simulates failure of the reference tube

After the supply voltage has decreased to zero, check for heating of components in the circuit, notably resistors. Most should run warm but not hot. The exception here is R5, which will tend to run hot depending on the setting of R9.

Under use, the regulator was observed to perform very well. Output stability was within expectations, and repeatability was good; that is, the output voltage of the supply returned to its set voltage from one period of operation to the next. The warm-up time to reach the set output voltage was observed to be about five minutes or so.

Final Power Supply Design

Using the basic RCA power supply/regulator design shown in Figures 8.1 and 8.2, and the experience gained in building and testing the circuits, the author arrived at a final general-purpose power supply design, as shown in Figure 8.6. The power supply and filter are conventional in approach. The voltage regulator follows closely the basic RCA circuit, with the exception of the changes detailed previously (improved protection in the event of a failure or removal of the 5651A tube and restricting the possible range of settings for the voltage adjustment potentiometer).

The 5BC3 power rectifier feeds a choke-input filter followed by a capacitor bank. Optionally, a capacitor-input filter may be used; the optional devices are C1 and C2, along with voltage-distributing resistors R1 and R2 and series resistor R24. The capacitor-input filter will provide a significant increase in output voltage at the cost of somewhat reduced regulation under varying loads. Table 8.6 summarizes the voltage output options. The maximum rated load of the power supply (unregulated + regulated loads) is 288 mA, set by the rating of transformer T1 (Hammond 300 BX).

The voltage regulator provides an adjustable output of +250 V DC to +275 V DC for preamplifier and ancillary circuits for an input voltage of approximately +400 V. At higher voltages, the range of adjustment (using the component values specified

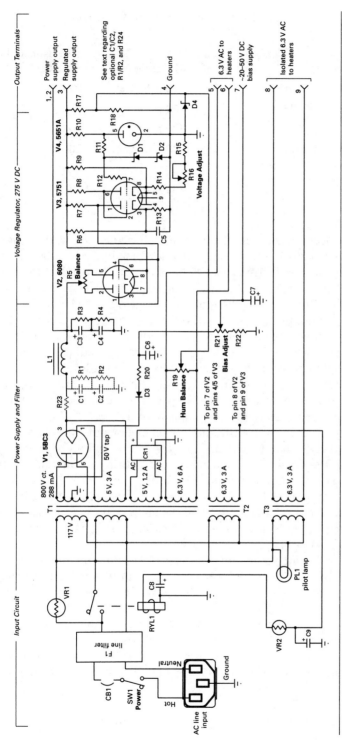

FIGURE 8.6 Final power supply design

TABLE 8.6 Power Supply Output Characteristics of the Final Design

Filter Option	Operating Load	Approximate Output Voltage
Capacitor-input filter configuration (C1/C2, R1/R2, and R24 installed)	120 mA	460
	160 mA	435
	200 mA	420
Choke-input filter configuration (C1/C2, R1/R2 not installed, jumper across R24)	120 mA	320
	160 mA	310
	200 mA	290

in Table 8.7) slides higher; for example, for an input voltage of +475 V, the range of adjustment is +270 V to +305 V DC. A voltage-divider/bleeder resistor ladder on the output of the regulator (R17 and R18) is used to develop approximately 40 V on the filament circuit to minimize hum. A zener diode protects the filament circuit in the event of a failure in the shunt resistor (R18) of the voltage divider ladder.

A grid bias supply of nominally –30 V is developed through a silicon rectifier and capacitor-input filter connected to the bias supply winding of the plate transformer. A potentiometer at the output of the supply is provided for fine adjustment of bias. Resistor R22 is included to prevent adjustment of bias to a lower level than might be advisable for the amplifier. Using the parts specified in Table 8.7, the range of adjustment is in the neighborhood of –25 to –50 V DC. If needed, the value of R22 may be changed to provide for more (or less) adjustment range.

The use of an oversized power transformer and choke ensure ample energy for power output stages. Bleeder resistors are included across each filter capacitor to load the supply and reduce the possibility of electrical shock during maintenance. The main filter capacitor banks consist of two 100 µF 450 V electrolytic capacitors in series (effectively ~50 µF). Resistors of 100 kΩ 2 W are placed in parallel with each capacitor to equalize the voltage distribution and bleed off the charge when power is removed. The series filter capacitors were chosen over higher-voltage single devices because of cost benefits and layout simplicity.

As shown in Figure 8.6, an AC line input voltage of 117 V nominal is assumed. The primary of plate transformer T1 can be connected to accommodate other voltages; however, the filament transformers (T2 and T3), as specified, have only 117 V primary windings.

Power thermistor VR1 is used to limit current inrush upon application of AC line voltage to the transformers. VR1 is rated for 8 A steady-state current. When "cold" (room temperature), the device exhibits a resistance of 2.5 Ω; however, when loaded to 50 percent of rated current, the resistance is just 0.14 Ω. The benefit of having VR1 in the primary lead is found only at power-up. During steady-state operation, the device performs no real purpose (other than generating heat). Furthermore, the resistance of the device, while small, can impact the B+ supply voltage under high power output due to the series resistance it adds to the primary circuit (0.5 V drop across VR1 has been observed). For this reason, relay RYL1 has been added to close the circuit around VR1 after the power supply has reached its operating point. Power for RYL1 is derived from one of the 5 V filament windings that are not otherwise used on the Hammond 300BX transformer. The 5 V AC input is rectified by a bridge, filtered, and applied to RYL1 through varistor VR2. Due to the time constants involved, RLY1 typically closes about one minute after power to the supply is switched on. The ambient temperature impacts the time-to-closure of RYL1, but is not critical to proper operation. A parts list for the general-purpose power supply shown in Figure 8.6 is given in Table 8.7.

The construction considerations outlined previously apply here as well.

TABLE 8.7   Electrical Components for the General-Purpose, High-Voltage Power Supply

Part	Description	Quantity	Manufacturer	Part No.	Allied Stock No.
C1, C2, C3, C4	100 µF, 450 V, electrolytic	4	Cornell-Dubilier	380LX101M450J022	862-0144
C5	0.1 µF, 600 V	1	Vishay	715P10456LD3	507-0328
C6, C7	22 µF, 100 V	2	Illinois Capacitor	226TTA100M	613-0352
C8, C9	1,000 µF, 25 V electrolytic	2	Illinois Capacitor	108TTA025M	613-0327
R1, R2, R3, R4	100 kΩ, 2 W	4	Ohmite	OY104KE	296-2322
R5	50 Ω, 25 W rheostat	1	Ohmite	RHS50RE	296-4113
R6, R9	12 kΩ, 2 W	2	Vishay	PR02000201202JR500	648-0160
R7, R8, R11, R12	470 kΩ, 0.5 W	4	Ohmite	OL4745E	296-4789
R10	68 kΩ, 2 W	1	Ohmite	OY683KE	296-2384
R13	15 kΩ, 2 W	1	Ohmite	OY153KE	296-2333
R14, R18	4.7 kΩ, 2 W	2	Ohmite	OY472KE	296-2369
R15	330 Ω, 2 W	1	Ohmite	OY331KE	296-2353
R16, R21	5 kΩ, 2 W potentiometer	2	Honeywell/Clarostat	RV4LAYSA502A	753-1145
R17	27 kΩ, 2 W	1	Ohmite	OY273KE	296-6377
R19	100 Ω, 2 W, potentiometer	1	Honeywell/Clarostat	RV4LAYSA101A	753-1001
R20	100 Ω, 2 W	1	Ohmite	OY101KE	296-2318
R22	2.7 kΩ, 2 W	1	Ohmite	OY272KE	296-2349

(continued)

TABLE 8.7   Electrical Components for the General-Purpose, High-Voltage Power Supply (*Continued*)

Part	Description	Quantity	Manufacturer	Part No.	Allied Stock No.
R23	50 Ω, 20 W	1	Ohmite	B20J50RE	296-6270
D1	Zener diode, 51 V, 5 W	1	ON Semiconductor	1N5369BG	568-0593
D2	Zener diode, 47 V, 5 W	1	ON Semiconductor	1N5368BG	568-0592
D3	Rectifier, 1 A, 600 V	1	NTE	116	935-0298
D4	Zener, 75 V, 5 W	1	ON Semiconductor	1N5374BG	568-0595
CR1	Bridge rectifier, 100 V, 2 A	1	NTE	NTE166	935-6256
PL1	117 V AC pilot lamp	1	Dialight	249-7841-1431-574	511-0276
L1	Choke: 5 H 300 mA 55 Ω	1	Hammond[1]	193LP	Newark[1]
T1	Power transformer: Primary– 120 V Secondary– 400-0-400 V rms, 288 mA 6.3 V, 5 A 5 V, 3 A 5 V, 1.2 A	1	Hammond	300BX	Newark
T2, T3	Filament transformer, 6.3 V, 6 A	2	Hammond	167Q6	Newark
VR1	Power thermistor, 2.5 Ω cold	1	GE	CL-30	837-0030
VR2	Thermistor, 20 Ω cold	1	Honeywell	ICL1220002-01	254-0094

(*continued*)

TABLE 8.7 Electrical Components for the General-Purpose, High-Voltage Power Supply (*Continued*)

Part	Description	Quantity	Manufacturer	Part No.	Allied Stock No.
F1	AC line filter	1	Qualtek	848-10/003	689-3564
SW1	SPST switch	1	Eaton	7504K4	757-4201
CB1	Circuit breaker, 8 A	1	Tyco	W58-XB1A4A-8	886-8876
RYL1	5 V DC relay, contacts 240 V AC, 10 A	1	Magnecraft	9AS1D52-5	850-0123
	4 terminal barrier strip	1	Molex	38720-6204	607-0071
V1	5BC3 tube	1			The Tube Store[2]
V2	6080 tube	1			Tube Depot[2]
V3	5751 tube	1	Sovtek, JAN Philips, others		Tube Depot
V4	5651A tube	1			Tube Store
	9-pin tube socket	1	Belton	P-ST9-601	AES[3]
	9-pin novar socket	1	Belton	SK-9PINX	Tube Depot
	7-pin tube socket	1	Belton	P-ST7-201MXB	AES
	Octal tube socket	1	Belton	P-ST8-209MIP	AES
	¾-inch #4 threaded standoff	12	Keystone Electronics	1895	839-0781
	AC power cord	1	Alpha Wire	615 BK078	663-7020

[1] Hammond Manufacturing Company, Inc., 475 Cayuga Rd., Cheektowaga, NY,14225, USA: www.hammondmfg.com. Available from Newark Electronics (www.newark.com).

[2] Vacuum tubes are available from a number of suppliers, including Tube Depot (www.tubedepot.com), The Tube Store (http://thetubestore.com), and Tube World (www.tubeworld.com).

[3] Available from Antique Electronic Supply (www.tubesandmore.com).

Circuit breaker CB1 protects the power supply from damage due to component failure or a short-circuit of the load(s). As specified in Table 8.7, the trip point of CB1 is 8 A. This device is available at a number of other operating currents. Readers should select a circuit breaker rated for the expected loading of the supply, but in any event not greater than 8 A. Since this power supply may draw a fair amount of current when working at high power levels, it should be connected to a source of AC power that is capable of handling the load without voltage sag due to loading or some other device on the circuit. A dedicated 15 A circuit would be ideal. The power supply can certainly share a 15 A circuit with a preamp, tuner, phono turntable, and related devices. The unit should not share an AC mains circuit with high-current loads such as heaters or other high-wattage appliances.

Note that when the optional capacitor-input filter is used (C1/C2, R1/R2, and R24 installed), under no-load conditions, the capacitor bank consisting of C1/C2 will charge up to 1.4 times the applied secondary voltage of T1, which is approximately +560 V DC. The components specified in Table 8.7 are rated for operation at this voltage (and higher). Ensure that any circuits connected to the unregulated output (terminals 1 and 2) can safely operate at this level. When power is first applied, the voltage at the input of the filter (the junction of C1 and L1) will rapidly rise to the peak value, and then as the filaments warm up and the load begins to draw current, the voltage will decrease to the steady-state operating value. It is during the warm-up period that the highest voltage appears across load devices, in particular, electrolytic capacitors. While the circuit-operating voltage for a particular electrolytic might be +300 V DC, there may be times when the device is exposed to the full no-load supply voltage. Devices should, therefore, be specified accordingly.

Electrolytic capacitors are rated by their *working voltage* and *surge voltage.* The surge voltage for a high-voltage capacitor (in the 450 V range) might be 50 volts above the working voltage. Exceeding the surge voltage rating may result in failure of the capacitor.

The voltage output of the regulated supply may overshoot the set voltage by a small amount during warm-up. After a few minutes of operation, the voltage will return to its set level as the regulator tube warms up. Other than this situation, the regulated output does not experience the large overshoot seen with the unregulated output.

PWB Design

For the final power supply design, a two-board PWB implementation was developed that incorporates most components, except for the transformers, primary AC line devices, and variable resistors. The overall component layout for the filter board is shown in Figure 8.7. The regulator board is shown in Figure 8.8. All leads for the secondary windings of transformers T1, T2, T3, and L1 terminate on the PWBs. The wire connection codes (using the transformers specified in Table 8.7) are given in Table 8.8.

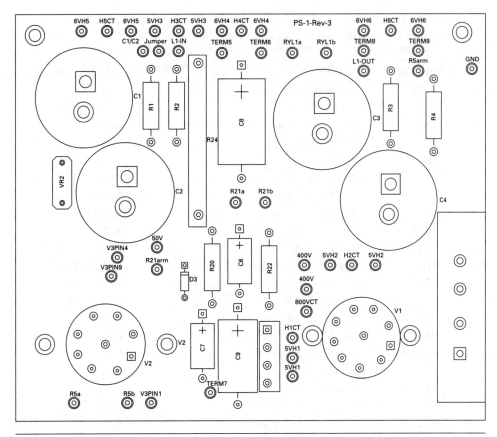

FIGURE 8.7 Component layout for the power supply rectifier PWB

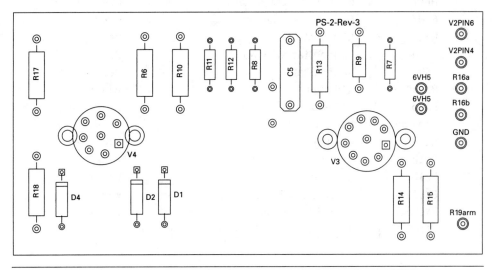

FIGURE 8.8 Component layout for the power supply regulator PWB

TABLE 8.8   PWB Solder-On Connection Wiring Codes. (Note that the color codes for the transformers listed in this table may vary. Consult the device manufacturer.)

PWB Code	Function	Connection Color Code	Device
400V	400 V AC, 288 mA, plate winding	Red	T1
800CT	400-400 V center-tap	Red/yellow	
400V	400 V AC, 288 mA, plate winding	Red	
50V	50 V AC bias tap	Violet	
5VH1	5 V AC, 1.2 A, heater	Black/yellow	
H1CT	5 V filament center-tap	Orange	
5VH1	5 V AC, 1.2 A, heater	Black/yellow	
5VH2	5 V AC, 3 A, heater	Yellow	
H2CT	5 V filament center-tap	Yellow/black	
5VH2	5 V AC, 3 A, heater	Yellow	
5VH3	5 V AC, 1.2 A, heater	Red/white	
H3CT	5 V heater center-tap	White/red	
5VH3	5 V AC, 1.2 A, heater	Red/white	
6VH4	6.3 V AC, 6 A, heater	Green	
H4CT	6.3 V filament center-tap	Green/Yellow	
6VH4	6.3 V AC, 6 A, heater	Green	
L1	Choke	Black	L1
L2	Choke	Black	
6VH5	6.3 V AC, 6 A, heater	Green	T2
H5CT	6.3 V AC heater center-tap	Green/yellow	
6VH5	6.3 V AC, 6 A, heater	Green	
6VH6	6.3 V AC, 6 A, heater	Green	T3
H6CT	6.3 V AC heater center-tap	Green/yellow	
6VH6	6.3 V AC, 6 A, heater	Green	
R5a	R2 rheostat	Red	R2
R5arm	R2 rheostat wiper arm	Red	
R5b	R2 rheostat	Red	
R16a	R16 high	White	R16
R16b	R16 wiper/low	White	
R19arm	R19 wiper	Green	R19
R21a	R21 high	White	R21
R21arm	R21 wiper	White	
R21b	R21 low	White	
RYL1a	Relay RYL1 winding	White	RYL1
RYL1b	Relay RYL1 winding	White	
C1/C2 Jumper	Jumper for capacitor-input configuration	Red	C1/C2

The PWBs are connected to the chassis through the mounting holes for the vacuum tubes. The tube sockets are conventional chassis-mounted devices, with jumper leads extending from the active pins to the PWB. The sockets are held away from the PWB using standoffs. This approach permits the sockets to be firmly mounted on the chassis, rather than being physically supported by the PWB. Furthermore, with this mounting arrangement, radiated heat from the tubes is dissipated above the chassis, preventing heat damage to the PWB over time.

As mentioned previously, the power supply may optionally use a capacitor-input filter. The PWB has been configured to include C1, C2, R1, and R2. These devices can be installed on the board and connected into (or disconnected from) the filter circuit by adding (or removing) the "C1/C2 Jumper" on the PWB. When the capacitor-input filter option is used, resistor R24 should be installed; otherwise, install a jumper in place of R24.

The two-board design is useful, in that each board can be used for different purposes. For example, the rectifier board can be used for a variety of power supplies. Likewise, the regulator board can be repurposed for other power supply applications. For most cases, however, it is assumed the rectifier and regulator boards will be used together. That being the case, a combined board is shown in Figure 8.9, on the following page. In this particular situation, the manufacturing cost of the PWB is lower for one large board than for two smaller ones.

It is not permissible (or practical, really) to reproduce the PWB circuit traces for the power supply boards in this book. The layout files are, however, available from the author at the VacuumTubeAudio.info website, as detailed at the end of the book.

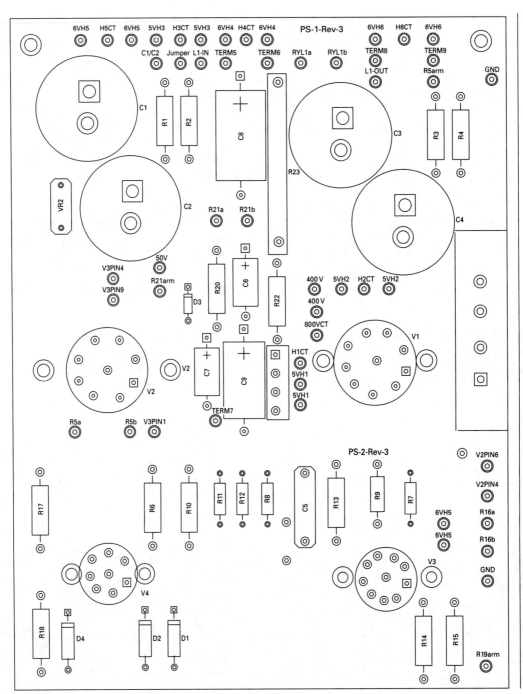

FIGURE 8.9 Power supply/regulator PWB component layout

Chapter 9

Project 2: Audio Preamplifiers

Project 2 is a high-fidelity audio preamplifier. The unit provides separate bass and treble controls and has three audio inputs:

- Phonograph (RIAA equalization)
- Microphone (high impedance)
- One line-level input

Design objectives include the following:

- Frequency response within ±3 dB, 20 Hz to 20 kHz
- Total harmonic distortion less than 0.5 percent
- Intermodulation distortion less than 0.25 percent
- Total hum and noise at least 65 dB below rated output (input terminals shorted)
- Total project cost for parts below $250

This project will be described as a series of subcircuits for the phono preamp, mic preamp, tone control, and buffer amplifier circuits.

Phonograph Preamp

The phono preamp is a two-stage, high-gain circuit intended for use with a magnetic phonograph pickup device [1] (see Figure 9.1). The 7025 twin triode features low hum and noise, and is designed specifically for use in high-fidelity circuits that operate at low signal levels. Typical voltage gain is in excess of 130.

The audio signal from the phono pickup is applied to the grid of V1a. Interstage coupling between the two amplifier sections includes an RIAA equalization network consisting of C3/R6, C6/R10, C7, and C8.[1] The output of the preamp is coupled from

---

[1] The RIAA equalization curve is described at the end of this chapter.

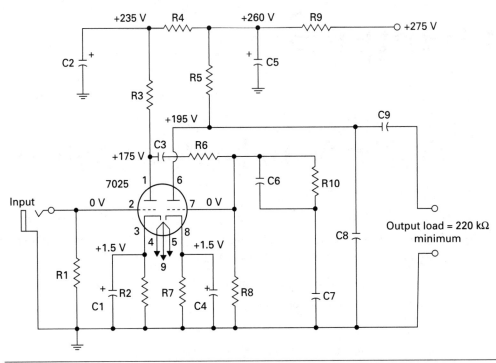

FIGURE 9.1   Schematic diagram of the phonograph preamplifier. Typical operating voltages are shown for reference. (*After* [1].)

the plate of the second stage by output coupling capacitor C9 to the input of the tone control amplifier (if used) or power amplifier.

Because of the relatively high output impedance, the preamp is recommended for use in systems in which the preamp is mounted on the same chassis as the power amplifier and/or tone control amplifier. The preamp may be used at distances of up to 6 ft from the power amplifier, provided the capacitance of C8 is reduced approximately 30 pF for each foot of shielded cable used for the audio connection between the preamp and the following amplifier. A minimum load of 220 kΩ is recommended.

Heater and DC power for the circuit can be obtained from the power supply for the audio power amplifier. The B+ supply should be in the range of 250 V DC to 300 V DC.

Parts List

Table 9.1 lists the electronic components needed to build this preamplifier, along with manufacturer part numbers and stock numbers from one of the major parts houses.[2]

---

[2] Allied Electronics, 7151 Jack Newell Blvd. S., Fort Worth, Texas, 76118, USA, www.alliedelec.com.

TABLE 9.1   Parts List for the Initial Build of the Phonograph Preamplifier

Part	Description	Quantity	Manufacturer	Part No.	Allied Stock No.
C1, C4	25 µF, 50 V, electrolytic	2	Vishay	TVA1306-E3	507-1089
C2, C5	22 µF, 450 V, electrolytic	2	Illinois Capacitor	226TTA450M	613-0367
C3	0.1 µF, 600 V	1	Vishay	715P10456LD3	507-0328
C6	0.0033 µF, 600 V	1	Vishay	715P33256JD3	507-0337
C7	0.01 µF, 600 V	1	Vishay	715P10356KD3	507-0326
C8	180 pF, 500 V, mica	1	Cornell-Dubilier	CD15FD181J03F	862-0547
C9	0.22 µF, 600 V	1	Vishay	715P22456MD3	507-0335
R1[1]	47 kΩ, 0.5 W	1	Ohmite	OL4735E	296-4788
R2, R7	2.7 kΩ, 0.5 W	2	Ohmite	OL2725E	296-6170
R3, R5	100 kΩ, 0.5 W	2	Ohmite	OL1045E	296-4773
R4	39 kΩ, 0.5 W	1	Ohmite	OL3935E	296-4784
R6	470 kΩ, 0.5 W	1	Ohmite	OL4745E	296-4789
R8	680 kΩ, 0.5 W	1	Tyco	LR1F680K	437-0431
R9	15 kΩ, 2 W	1	Ohmite	OY153KE	296-2333
R10	22 kΩ, 0.5 W	1	Ohmite	OL2235E	296-4779
V1	7025 tube	1	Electro-Harmonix, Genalex, others		Tube Depot[2]

[1] Optimum value depends on the type of magnetic pickup used. Follow manufacturer's recommendations.

[2] Vacuum tubes are available from a number of suppliers, including Tube Depot (www.tubedepot.com), The Tube Store (http://thetubestore.com), Tube World (www.tubeworld.com), and other suppliers.

These parts are, of course, available from a number of manufacturers and suppliers. The details given here are for the convenience of the reader.

Components for the power supply are not shown, as it is assumed this preamp will use the power supply described in Chapter 8 or be built into a power amplifier, such as the one described in Chapter 10 or Chapter 11.

For a stereo preamplifier, the quantities listed in the table should be doubled. Considerations relating to chassis components and assembly techniques are addressed in other chapters.

At the time of this writing, the 7025 tube was being manufactured by one or more tube companies.

Microphone Preamp

This single-stage preamp is intended for use with high-impedance dynamic microphones [1] (see Figure 9.2). The circuit uses a 5879 low-noise sharp-cutoff pentode in a conventional amplifier circuit with a high-impedance output and high voltage gain (approximately 60). Because of the high output impedance, the preamp should be mounted on the same chassis as the power amplifier and tone-control amplifier (if used). A minimum load of approximately 220 kΩ is recommended.

Heater and DC power for the circuit can be obtained from the power supply for the audio power amplifier. The B+ supply should be in the range of 250 V DC to 300 V DC.

Parts List

Table 9.2 lists the electronic components needed to build this preamplifier, along with manufacturer part numbers and stock numbers. Components for the power supply are not shown. For a stereo preamplifier, the quantities listed in the table should be doubled.

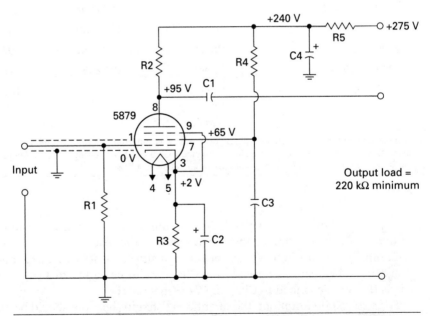

FIGURE 9.2 Schematic diagram of the microphone preamplifier. Typical operating voltages are shown for reference. (*After* [1].)

TABLE 9.2   Parts List for the Initial Build of the Microphone Preamplifier

Part	Description	Quantity	Manufacturer	Part No.	Allied Stock No.
C1	0.047 µF, 400 V	1	Vishay	715P47354LD3	507-0342
C2	25 µF, 50 V, electrolytic	1	Vishay	TVA1306-E3	507-1089
C3	0.22 µF, 600 V	1	Vishay	715P22456MD3	507-0335
C4	47 µF, 450 V, electrolytic	1	Illinois Capacitor	476TTA450M	613-0369
R1	2.2 mΩ, 0.5 W	1	Ohmite	OL2255E	296-6466
R2	100 kΩ, 0.5 W	1	Ohmite	OL1045E	296-4773
R3	1 kΩ, 0.5 W	1	Ohmite	OL1025E	296-4772
R4	470 kΩ, 0.5 W	1	Ohmite	OL4745E	296-4789
R5	22 kΩ, 0.5 W	1	Ohmite	OL2235E	296-4779
V1	5879 tube	1	NOS		Tube Depot[1]

[1] Vacuum tubes are available from a number of suppliers, including Tube Depot (www.tubedepot.com), The Tube Store (http://thetubestore.com), Tube World (www.tubeworld.com), and other suppliers.

At the time of this writing, the 5879 tube was available as new old stock (NOS) from one or more tube suppliers.

Tone Control Amplifier

This high-fidelity tone control amplifier uses a 6EU7 low-noise twin triode in a two-stage cascade that consists of an input cathode-follower connected to a triode voltage amplifier through a frequency-sensitive (tone control) interstate coupling network [1] (see Figure 9.3). The bass and treble controls in the coupling network can be adjusted to provide for up to approximately 16 dB of boost or attenuation (cut) at 30 Hz and 15 kHz.[3] With the bass and treble controls set at the midrange positions, the amplifier provides a reasonably flat response curve and a nominal voltage gain of 3.

The tone control amplifier is designed for use immediately ahead of an audio power amplifier. For operating convenience, the volume control for the power amplifier may be physically located on the tone control chassis. In this case, it is advisable to insert a 1 mΩ resistor in place of the volume control on the power amplifier. A minimum load of 100 kΩ is recommended.

Heater and DC power for the circuit can be obtained from the power supply for the audio power amplifier. The B+ supply should be in the range of 250 V DC to 300 V DC.

---

[3] See Table 9.9 for actual measured values.

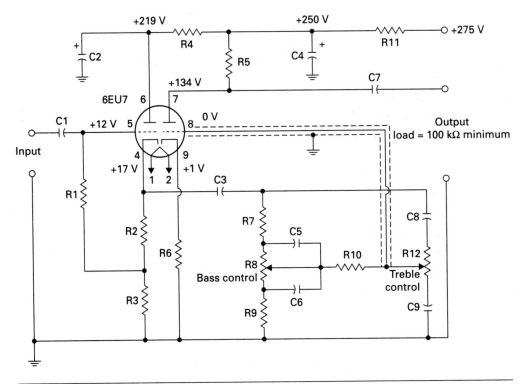

FIGURE 9.3 Schematic diagram of the tone control preamplifier. Typical operating voltages are shown for reference. (*After* [1].)

Parts List

Table 9.3 lists the electronic components needed to build the tone control preamplifier, along with manufacturer part numbers and stock. For a stereo amplifier, the quantities listed in the table should be doubled.

At the time of this writing, the 6EU7 tube was being manufactured by one or more tube companies.

Buffer Amplifier

This two-stage buffer amplifier is used to transform the high-impedance outputs of the previous preamplifier circuits to a lower impedance that is less susceptible to capacitive loading from interconnecting cables, thereby simplifying connection to external devices (see Figure 9.4). The first stage of the 7025/12AX7 is a conventional voltage amplifier, with the plate connected directly to the grid of the second stage. The cathode-follower output is taken across a 100 kΩ load. Voltage gain is nominally 30.

TABLE 9.3   Parts List for the Initial Build of the Tone Control Preamplifier

Part	Description	Quantity	Manufacturer	Part No.	Allied Stock No.
C1	0.047 μF, 400 V	1	Vishay	715P47354LD3	507-0342
C2, C4	22 μF, 450 V, electrolytic	2	Illinois Capacitor	226TTA450M	613-0367
C3	0.1 μF, 600 V	1	Vishay	715P10456LD3	507-0328
C5, C9	0.0022 μF, 600 V	1	Vishay	715P22256JD3	507-0332
C6	0.022 μF, 600 V	1	Vishay	715P22356KD3	507-0334
C7	0.22 μF, 600 V	1	Vishay	715P22456MD3	507-0335
C8	220 pF, 600 V, ceramic	1	Vishay	562R5GAT22	507-0815
R1	470 kΩ, 0.5 W	1	Ohmite	OL4745E	296-4789
R2	1.5 kΩ, 0.5 W	1	Ohmite	OL1525E	296-4774
R3, R11	15 kΩ, 0.5 W	2	Ohmite	OL1535E	296-4775
R4	22 kΩ, 0.5 W	1	Ohmite	OL2235E	296-4779
R5, R7, R10	100 kΩ, 0.5 W	3	Ohmite	OL1045E	296-4773
R6	1 kΩ, 0.5 W	1	Ohmite	OL1025E	296-4772
R8, R12	1 mΩ potentiometer, audio taper	2	Precision Electronic Components[1]	KA1051S28-ND	
R9	10 kΩ, 0.5 W	1	Ohmite	OL1035E	296-1483
V1	6EU7 tube	1	Sovtek, others		Tube Depot[2]

[1] Available from Digi-Key Corp (www.digikey.com).
[2] Vacuum tubes are available from a number of suppliers, including Tube Depot (www.tubedepot.com), The Tube Store (http://thetubestore.com), Tube World (www.tubeworld.com), and other suppliers.

For certain applications, complete isolation of the output from the load may be preferable or necessary. For such situations, a high-impedance transformer (primary impedance of 80 kΩ or so) and a low impedance balanced secondary (600 Ω is common) may be used. If you go this route, make sure to locate the transformer after the output blocking capacitor, as many devices of this class are intended for low-level operation and are not rated for high DC voltages on their windings. These devices can be hard to find and tend to be quite expensive.

This circuit is designed to be as flat and noiseless as possible, as its primary purpose is isolation and interface.

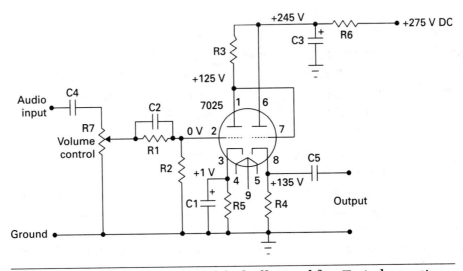

FIGURE 9.4   Schematic diagram of the buffer amplifier. Typical operating voltages are shown for reference.

Parts List

Table 9.4 lists the electronic components needed to build the buffer amplifier, along with manufacturer part numbers and stock numbers. For a stereo amplifier, the quantities listed in the table should be doubled.

TABLE 9.4   Parts List for the Initial Build of the Buffer Amplifier

Part	Description	Quantity	Manufacturer	Part No.	Allied Stock No.
C1	47 µF, 50 V, electrolytic	1	Illinois Capacitor	476TTA050M	613-0342
C2	100 pF, 1000 V disc.	1	Vishay	561R10TCCT10	507-0889
C3	22 µF, 450 V	1	Illinois Capacitor	226TTA450M	613-0367
C4, C5	0.22 µF, 400 V	2	Vishay	715P22454MD3	507-0664
R1, R2	220 kΩ, 0.5 W	2	Ohmite	OL2245E	296-4780
R3, R4	100 kΩ, 0.5 W	2	Ohmite	OL1045E	296-4773
R5	1 kΩ, 0.5 W	1	Ohmite	OL1025E	296-4772
R6	15 kΩ, 2 W	1	Ohmite	OY153KE	296-2333
R7	1 MΩ potentiometer, audio taper	1	Precision Electronic Components[1]	KA1051S28-ND	
V1	7025/12AX7 tube	1	Svetlana, others		Tube Depot[2]

[1] Available from Digi-Key Corp (www.digikey.com).
[2] Vacuum tubes are available from a number of suppliers, including Tube Depot (www.tubedepot.com), The Tube Store (http://thetubestore.com), Tube World (www.tubeworld.com), and other suppliers.

At the time of this writing, the 7025/12AX7 tube was being manufactured by one or more tube companies.

Construction Considerations

The audio preamplifier circuits described in this chapter can be constructed in any number of ways. For example purposes, this unit was hand-wired using terminal strips on a medium-size aluminum chassis. A parts list for the chassis components is given in Table 9.5. Note that power supply components are not included, as it is assumed the heater and B+ voltages are provided by a separate power supply or power amplifier. In this implementation, power was supplied by the regulated power supply described in Chapter 8. Bulk supplies needed for project assembly are listed in Chapter 6 (Table 6.6).

The construction of the amplifier is conventional in approach. The completed unit is documented in Figure 9.5.

After completing the preamplifier circuits, some general observations can be made, including the following:

- The chassis chosen (see Table 9.5) is just large enough to build this project with all four circuits described in this chapter. It can be seen from the photographs in Figure 9.5 that component spacing is rather tight, but workable. As expected, placement of capacitors presented the greatest challenge. Because of the tight spacing, component leads were covered with "spaghetti" in most cases to reduce the possibility of short circuits.
- Two rows of terminal strips were placed on either side of the four tube sockets for a total of 8 five-contact positions. To the extent possible, input circuit connections were located on one side of the chassis and output circuit connections were located on the other side.

TABLE 9.5   Chassis Parts List for the Initial Build Circuits

Description	Quantity	Manufacturer	Part No.	Allied Stock No.	Notes
Chassis, 9.5-in × 5-in × 2.5-in	1	Bud	CU-592-A	736-3611	
9-pin tube socket	4	Belton	SK-B-VT9-ST-2		Tube Depot
Knob	3	Davies	1110	543-1110	Volume control
Chassis feet	4	Bud	F-7264-A	736-7264	
RCA connector	4	Neutrix	NF2D-B-0	514-0004	Audio connector
Rotary switch	1	Electroswitch	C4D0206N-A	747-6690	Phono/mic/aux selector
Terminal strip	8	Cinch	54A	750-6628	

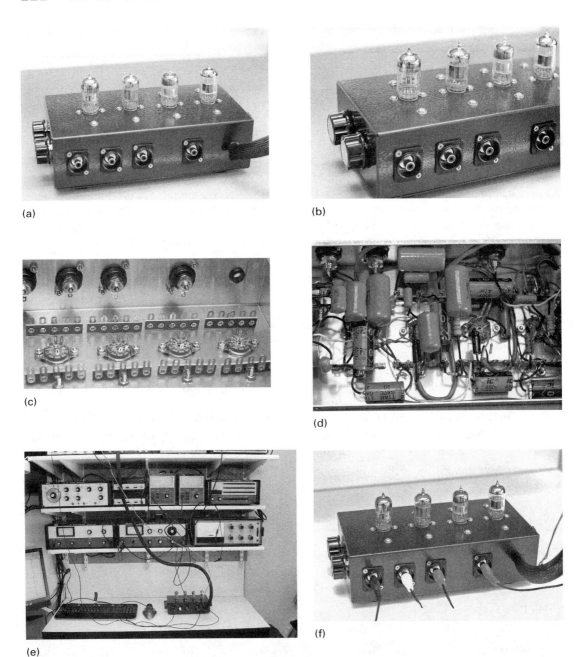

FIGURE 9.5 The audio preamplifier: (*a*) top view, (*b*) close-up view, (*c*) bottom view prior to installation of components, (*d*) bottom view with components installed, (*e*) completed preamp under bench test, (*f*) unit under test—note the shorting input connectors for noise measurements.

- Controlling hum on the low-level mic and phono circuits was a considerable challenge. Distortion measurements were limited by the noise floor, which was largely dominated by 60 Hz hum. While acceptable performance can be achieved with this physical layout, a better approach would be to place the low-level stages (mic and phono) in separate self-contained chassis; for example, the Hammond 1590WE (Allied stock #806-5907) diecast aluminum enclosure (approximately seven inches by five inches by three inches deep). This would give greater control over noise sources. The larger available chassis space would also permit greater separation of components, making for easier construction and rework if necessary. Building the circuits on a printed wiring board with large ground plane areas is another viable approach. Shielded tube sockets are also recommended for the mic and phono circuits, as sensitivity to the local environment was observed during proof of performance measurements.

Initial Checkout

After the preamplifier has been assembled, the following steps are recommended to confirm proper construction:

1. Remove all power from the unit.
2. Remove all tubes.
3. Using an ohmmeter, confirm that the resistance values listed in Table 9.6 are observed on the tube socket connections. All readings are with reference to ground. Readings within ±5 percent are considered normal. The resistance measurements are made with the power supply disconnected.

TABLE 9.6   Typical Ohmmeter Readings with Power Removed and Tubes Removed

Tube	Pin No.	Typical Reading	Notes
7025, Phono Preamp	1	> 200 kΩ	Varies due to filter capacitor in circuit.
	2	47 kΩ	Measured with no input connected.
	3	2.7 kΩ	
	4	–	Heater
	5	–	Heater
	6	> 200 kΩ	Varies due to filter capacitor in circuit.
	7	680 kΩ	
	8	2.7 kΩ	
	9	–	Heater

(*continued*)

TABLE 9.6   Typical Ohmmeter Readings with Power Removed and Tubes Removed (*Continued*)

Tube	Pin No.	Typical Reading	Notes
5879, Microphone Preamp	1	2.2 mΩ	
	2	–	
	3	1 kΩ	
	4	–	Heater
	5	–	Heater
	6	–	
	7	> 1 mΩ	Varies due to filter capacitor in circuit.
	8	> 1 mΩ	Varies due to filter capacitor in circuit.
	9	1 kΩ	
6EU7, Tone Control Preamp	1	–	Heater
	2	–	Heater
	3	–	
	4	16.5 kΩ	
	5	480 kΩ	
	6	> 1 mΩ	Varies due to filter capacitor in circuit.
	7	> 1 mΩ	Varies due to filter capacitor in circuit.
	8	110 kΩ to 1.1 mΩ	Reading depends on setting of bass tone control. Values shown are for extreme positions.
	9	1 kΩ	
7025, Buffer Amplifier	1	> 1 mΩ	Varies due to filter capacitor in circuit.
	2	108 kΩ to 180 kΩ	Varies depending on setting of input volume control.
	3	1 kΩ	
	4	–	Heater
	5	–	Heater
	6	> 1 mΩ	Varies due to filter capacitor in circuit.
	7	> 1 mΩ	Varies due to filter capacitor in circuit.
	8	100 kΩ	
	9	–	Heater

4. If any deviations from the expected values listed in Table 9.6 are observed, recheck wiring and components for correct installation. Do not advance to the next step until the resistance measurements are within tolerance.

5. After the resistance checks have been successfully completed, connect the power supply and apply power to the unit. Be very careful, since high voltages are present.

6. Using extreme caution, check for proper voltages at the tube connection points listed in Table 9.7. All voltages are measured with respect to ground (unless otherwise noted) using a high-impedance voltmeter. Variations within ±10 percent

TABLE 9.7   Typical Voltmeter Readings with Tubes Removed and Power Applied[1]

Tube	Pin No.	Typical Reading	Notes
7025, Phono Preamp	1	272 V	
	2	0 V	No input connected.
	3	0 V	
	4	–	Heater
	5	–	Heater. There should be 0 V between pins 4 and 5.
	6	276 V	
	7	0 V	
	8	0 V	
	9	–	Heater. There should be 6.3 VAC between pin 9 and pins 4 and 5.
5878, Microphone Preamp	1	0 V	
	2	0 V	
	3	0 V	
	4	–	Heater
	5	–	Heater. There should be 6.3 VAC between pins 4 and 5.
	6	0 V	
	7	265 V	
	8	275 V	
	9	0 V	
6EU7, Tone Control Preamp	1	–	Heater
	2	–	Heater. There should be 6.3 VAC between pins 1 and 2
	3	0 V	
	4	0 V	
	5	0 V	
	6	268 V	
	7	273 V	
	8	0 V	
	9	0 V	
7025, Buffer Amplifier	1	270 V	
	2	0 V	
	3	0 V	
	4	–	Heater
	5	–	Heater. There should be 0 V between pins 4 and 5.
	6	275 V	
	7	270 V	
	8	0 V	
	9	–	Heater. There should be 6.3 VAC between pins 4 and 9 and between pins 5 and 9.

[1] A regulated B+ supply of 275 V DC was provided for powered measurements, using the power supply circuit described in Chapter 8.

are normal. These measurements can be taken from the top of the chassis by inserting the voltmeter probe into the socket connection points. This is safer than attempting to make the measurements from underneath the chassis.

7. If variations in the expected voltage readings are found, remove all power from the unit and recheck components and wiring. Do not advance to the next step unless the voltages check correctly.

8. Remove AC power from the unit.

9. Install all tubes.

10. Set the tone controls to their center positions.

11. Connect an audio amplifier to the output terminal.

12. Apply power and observe the tubes for any signs of overheating or other unexpected behavior. The filaments should be lit. Remove power if any problems are detected, or if noticeable hum can be heard from the speaker. Recheck wiring and components as needed.

13. Typical operating voltages are given on the schematic diagram for each stage. Exercise extreme caution in making any measurements on operating circuits.

14. After these preliminary tests have been successfully completed, proof of performance measurements should be taken on the preamplifier circuits as detailed in the next section.

Measured Performance

Upon completion of the preamplifier circuits, measurements were taken on each stage at an output level of 1 V rms into a load of 1 MΩ, except for measurement of the *maximum operating point,* as noted in Table 9.8. The following test equipment was used:[4]

- Heathkit IG-18 audio generator
- Heathkit IG-1275 lin/log sweep generator
- Heathkit IM-5258 harmonic distortion analyzer
- Heathtkit SM-5248 intermodulation distortion analyzer
- Heathkit IM-5238 AC voltmeter
- Heathkit IM-4100 frequency counter
- Heathkit IO-4510 oscilloscope

The measured performance of the preamplifier circuits is detailed in Table 9.8.

For measurement of the maximum operating point, the outputs of the mic, phono, tone control, and buffer stages were selected individually and applied to the test equipment. For measurements at the operating output level of 1 V rms, the outputs

---

[4] The author readily acknowledges that far more advanced test equipment is available today. Most of the measurements taken here can be accomplished with a computer-driven instrument or software running on a PC. The 1970s-era test equipment was assembled not by accident or default, but by design. Each piece was acquired, completely refurbished, and recalibrated to meet new equipment specifications. The whole idea of building vacuum tube audio gear is to capture a particular sound, and how better to measure that than with test instruments of the time?

TABLE 9.8   Measured Performance of the Audio Preamplifier Circuits

Test	Parameter	Measured Value	Notes
B+ voltage input		275 V DC	Regulated supply
Buffer Preamplifier			
Operating point	Input level	0.032 V rms	For 1 V rms output at 1 kHz into 1 MΩ
Frequency response	Referenced to 1 kHz	10 Hz to 70 kHz, ±1 dB	Note 1
Distortion	THD	>0.5%	Note 2
	IMD	0.12%	4:1 mix ratio
Noise	Unweighted (shorted input terminals)	−68 dB	Note 3
Maximum operating point, 1 kHz, >3% THD	Input level	0.52V rms	Note 4
	Output of stage into 1 MΩ	14.5 V rms	
	Noise, unweighted (shorted input terminals)	−94 dB	
Tone Control Preamp			
Tone controls set for flattest possible response	Input level	0.30 V rms	For output of 1 V rms at 1 kHz into 1 MΩ
	Frequency response	±2 dB, 15 Hz to 28 kHz	Notes 5, 6
	THD	>0.5%	Note 2
	IMD	0.35%	4:1 mix ratio
	Noise, unweighted (shorted input terminals)	−71 dB	Note 3
Treble control extreme settings	Maximum boost	+13 dB at 20 kHz	Note 7
	Maximum cut	−22 dB at 20 kHz	
Bass control extreme settings	Maximum boost	+16 dB	Note 8
	Maximum cut	−20 dB	
Maximum operating point, 1 kHz, >3% THD	Input level	6.2 V rms	Note 4
	Output of stage into 1 MΩ load	19 V rms	
	Noise, unweighted (shorted input terminals)	−97 dB	
Phono Preamp			
Operating point	Input level	0.0072 V rms	For output of 1 V rms at 1 kHz into 1 MΩ

(continued)

TABLE 9.8 Measured Performance of the Audio Preamplifier Circuits (*Continued*)

Test	Parameter	Measured Value	Notes
Phono Preamp			
Frequency response	Relative to RIAA curve (see Figure 9.6)	±1 dB, 20 Hz to 20 kHz	See Table 9.9 and note 9
Distortion	THD	>0.8%	Notes 2, 10
	IMD	–	Note 11
Noise	Unweighted (shorted input terminals)	–56 dB	Notes 3, 12
Maximum operating point, 1 kHz >3% THD	Input level	0.062 V rms	Note 4
	Output of stage into 1 MΩ load	8.2 V rms	
	Noise, unweighted (shorted input terminals)	–72 dB	
Microphone Preamp			
Operating point	Input level	0.0145 V rms	For output of 1 V rms at 1 kHz into 1 MΩ
Frequency response	Relative to 1 kHz	±3 dB, 20 Hz to 46 kHz	Note 13
Distortion	THD	>0.5%	Note 2
	IMD	0.25%	4:1 mix ratio
Noise	Unweighted (shorted input terminals)	–58 dB	Notes 3, 12
Maximum operating point, 1 kHz, >3% THD	Input level	0.27 V rms	Note 4
	Output of stage into 1 MΩ load	17 V rms	
	Noise, unweighted (shorted input terminals)	–84 dB	

Notes:

1 The frequency response of the buffer amplifier was exceptionally flat. Response was down 1.5 dB (relative to 1 kHz) at 7 Hz on the low end and 100 kHz on the high end.

2 Total harmonic distortion (THD) measurements were taken at 20 Hz, 100 Hz, 1 kHz, 10 kHz, and 20 kHz unless otherwise noted. The worst (highest) measurement is the one quoted.

3 Noise measurements were taken with all covers in place. This can make a substantial difference in the reading. For example, the buffer amplifier noise performance improved 15 dB when the bottom cover plate was added.

4 Measurements were taken on all four preamp circuits at levels just below clipping. In each case, the clipping point was identified visually on an oscilloscope and the input level was then backed off just enough to bring the THD to less than 3 percent. This was deemed to be the maximum operating point of the circuit.

5 This measurement is best made with a sweep generator. First, adjust the treble tone control for best (flattest) response from 1 kHz to 20 kHz. The high end was found to peak at about 3.5 kHz, after which response fell off smoothly. Next, adjust the bass tone control for best response from 20 Hz to 1 kHz. The low end was found to peak at about 20 Hz. A dip of about 2 dB (relative to the 1 kHz reference) was observed at approximately 230 Hz.

6 After the optimum settings for the bass and treble controls have been found, adjust the knob set-screws so that the marking for each knob is pointing straight up (or some other reference orientation). This will denote the "flat" position for the controls.

TABLE 9.8   Measured Performance of the Audio Preamplifier Circuits (*Continued*)

7 Response with maximum treble boost was found to peak at 15 kHz (+14 dB), decreasing smoothly from that point with increasing frequency.

8 Response with maximum bass boost increased smoothly over the measured range, with the exception of a small dip of about 1 dB at approximately 650 Hz.

9 The overall frequency response of the phono preamplifier relative to the RIAA curve was disappointing in the initial build of this circuit. As constructed per Figure 9.1, the circuit performed outside the target ±3 dB tolerance, increasing to nearly 4 dB deviation from the target at 20 kHz. These measurements are documented in Table 9.9. There were two basic approaches to addressing this problem: 1) disassemble the circuit and measure the values of the frequency response–shaping network components to determine how they compare with the specified values, or 2) pad out the circuit as needed to address the falloff. The author took approach #2 (for the time being, at least). A small value capacitor (25 pF) was added across R6 and the measurements were repeated. The response with the padding in place is shown in the last column in Table 9.9. All measurements were repeated, but there were essentially no differences for any of the measurements below 1 kHz, and so only one set of numbers is listed in the table for 20 Hz to 1 kHz. It can be seen from the measurements that the previous response performance of ±4 dB, 20 Hz to 20 kHz, has been improved to ±1 dB, 20 Hz to 20 kHz. It would probably have been possible to narrow the deviation from target to less than 1 dB, but that's a project for another day.

10 By necessity, THD measurements on the phono preamp took into consideration the RIAA frequency response curve. If the same input level was used for all measurements, the preamp would overload on the low end and be buried in the noise at the high end. The following procedure was used: 1) a reference output level from the preamp was established at 1 kHz; 2) for measurements at 20 Hz, 100 Hz, and 200 Hz, the input level was adjusted so that the output from the preamp matched the level at 1 kHz; 3) for the measurement at 10 kHz, the input level was maintained at the 1 kHz level and the gain of the distortion analyzer was increased to compensate for the response rolloff; and 4) for the measurement at 20 kHz, the gain of the analyzer was fixed at the position for the 10 kHz measurement and the input level was increased to satisfy the reference input level of the instrument (about 9 dB increase).

11 No intermodulation distortion (IMD) measurements were taken on the phono preamp, as the frequency response curve of the RIAA transfer characteristic would invalidate the measurement.

12 The noise levels of the phono and mic preamps at the stated reference point were worse than expected. As noted previously, a larger physical chassis with better shielding would have likely improved the noise floor. Likewise, a shielded socket is advisable. In the versions tested, shielded sockets were not used.

13 The high-frequency response of the microphone preamp was exceptional, down 3 dB at 46 kHz. The circuit measured ±1 dB from 40 Hz to 20 kHz. For reference, the specification sheet for a Shure SM58 microphone (a popular professional product) quotes response for 50 Hz to 15 kHz. Response of the microphone falls rapidly at those extremes.

of the mic, phono, and tone control circuits were applied sequentially to the buffer preamplifier and measurements were taken from the output of the buffer. This was done in order to minimize capacitive loading of the high-impedance outputs and resulting rolloff in high-frequency response. As shown in Table 9.8, the buffer preamplifier is essentially transparent to the applied signal. Use of the buffer preamp provided considerable flexibility in test instrumentation, allowing the output signal to be daisy-chained to four or more instruments without degrading the high-end performance.

Measurements taken at the output of the buffer preamp showed the loading of the test instruments to have negligible impact on high-end frequency response. At 100 kHz, the addition of all test gear used in the measurements (audio voltmeter, harmonic distortion analyzer, intermodulation distortion analyzer, oscilloscope, and frequency counter) reduced the output relative to 1 kHz by less than 1 dB. In the range of 20 Hz to 20 kHz, connection of the test equipment to the output of the buffer had no discernable effect.

Note that input signal level and noise measurements for the mic, phono, and tone control circuits were taken directly at the output of each preamp stage in order to fully characterize the circuit under test.

RIAA Equalization

The purpose of pre-emphasis in disk recording and the complementary de-emphasis in playback is to boost the amplitude of the recorded waveform at higher frequencies to improve the signal-to-noise ratio (SNR) [2]. The emphasis, adopted as a standard by the Recording Industry Association of America (RIAA), is known as the *RIAA characteristic*. The complementary recording and reproduction curves are illustrated in Figure 9.6. Note the response is not a simple 6 dB per octave filter, but rather a shelved response, which was intended to compensate for certain constraints relating to fine-groove disk record production and the legacy practices that existed at the time the curve was adopted by the RIAA.

Measured Performance

The frequency response performance of the phono preamplifier listed in Table 9.8 is a composite of the performance of the circuit based on the measurements documented in Table 9.9. The table lists the response targets for specified frequencies from 20 Hz to 20 kHz. As detailed in Table 9.8 Note 9, two sets of measurements were taken on the phono preamp circuit. The values shown in the Notes column document the high-frequency response with the circuit padded to more closely match the RIAA curve.

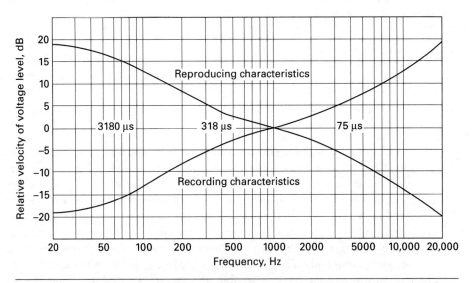

FIGURE 9.6 Characteristics for fine-groove records. (*RIAA Dimensional Characteristics for 33-1/3 rpm Records.*)

TABLE 9.9   Frequency Response Measurements as a Function of the RIAA Target Values

Frequency, Hz	RIAA Curve Target Level, dB	Measured Value	Notes
20	+19.3	+19.2 dB	
30	+18.6	+18.6 dB	
40	+17.8	+18.1 dB	
50	+17.0	+17.2 dB	
60	+ 16.1	+16.4 dB	
70	+ 15.3	+15.3 dB	
80	+14.5	+14.7 dB	
100	+13.1	+13.1 dB	
110	+ 12.4	+12.4 dB	
115	+11.6	+12.2 dB	
150	+10.2	+10.2 dB	
200	+8.3	+8.7 dB	
250	+6.7	+6.9 dB	
300	+ 5.5	+5.5 dB	
400	+3.8	+3.9 dB	
500	+ 2.6	+2.7 dB	
600	+ 1.9	+2.2 dB	
700	+ 1.2	+1.25 dB	
800	+ 0.7	+0.75 dB	
1,000	0	0 dB	Reference point
1,500	−1.4	−1.4 dB	−1.4 dB
2,000	−2.6	−2.5 dB	−2.6 dB
3,000	−4.7	−4.8 dB	−5.0 dB
4,000	−6.6	−6.8 dB	−6.7 dB
5,000	−8.2	−8.4 dB	−8.2 dB
6,000	−9.6	−10.2 dB	−9.6 dB
7,000	−10.7	−11.5 dB	−10.2 dB
8,000	−11.9	−12.8 dB	−12.0 dB
9,000	−12.9	−14.2 dB	−12.9 dB
10,000	−13.7	−15.5 dB	−13.8 dB
12,000	−15.3	−17.0 dB	−14.3 dB
14,000	−16.6	−19.2 dB	−15.6 dB
15,000	−17.2	−20.2 dB	−16.2 dB
16,000	−17.7	−21.5 dB	−16.8 dB
18,000	−18.7	−22.7 dB	−17.8 dB
20,000	−19.6	−23.5 dB	−18.8 dB

Final Preamplifier Design

Using the basic RCA circuits shown in Figures 9.1, 9.2, and 9.3, and the experience gained in building and testing those circuits, the author arrived at a final preamplifier design, as shown in Figure 9.7. The basic mic, phono, and tone control circuits follow closely the RCA design, with a few exceptions, as noted in the following text.

The phono and mic preamps are identical to the as-built designs in Figures 9.1 and 9.2, with the exception of output level controls R16 and R17. These devices, along with R18 and R19, are intended to be internally accessible gain controls used to even the levels from the four input sources so that adjustment of the main volume control is not necessary when switching from one input to the next. The switching is done at SW1, located just before the tone control stage. The four potentiometers are identical (1 mΩ). As specified in Table 9.10, the controls have a linear taper. Audio taper would work fine as well, of course.

Adjustment of R16 through R19 can be optimized through the following procedure:

1. Set the main volume control for approximately one-third clockwise rotation.
2. Set the bass and treble controls for flat response.
3. Identify the input source that will typically provide the lowest level; use this source as the reference.
4. Adjust the reference source gain control to yield the desired output from the buffer amplifier, such as 1 V rms.
5. Switch to each input and with typical program material applied, adjust the gain control of each input to provide approximately the same output level from the buffer stage as the reference. During this process, do not adjust the main volume control.

This procedure should provide a reasonable approximation of the proper settings. Fine-tuning may be necessary after additional use.

All audio cables of length greater than about six inches should use shielded cable. Generally speaking, the shield should be connected to ground at the load end. Grounding both ends of a shielded cable to the same ground plane is not recommended, as it may lead to circulating currents that will adversely impact the noise floor.

Capacitor CP in the phono preamp circuit may be needed to adjust the high-frequency response to more closely match the RIAA curve. As discussed in Table 9.8 Note 9, the test bed unit used a 25 pF capacitor. Table 9.10 lists this value in the parts list; however, capacitor CP is actually a "select on test" device. If the high-end frequency response is within 2 dB or so of the RIAA curve, capacitor CP does not need to be installed.

Potentiometer R39 serves as a "hum balance" control for the preamplifier. The potentiometer should be adjusted for minimum hum when the microphone input is selected (input terminals shorted). Note that R39 should not be installed if some other form of hum balance is implemented on the same filament transformer winding. For example, the power amplifier described in Chapter 10 applies a positive DC voltage (approximately 50 V) to the wiper arm of a hum balance control. Also, if a stereo unit is being constructed, use only one hum balance control, as adjustment of the controls could effectively short-circuit the filament transformer windings.

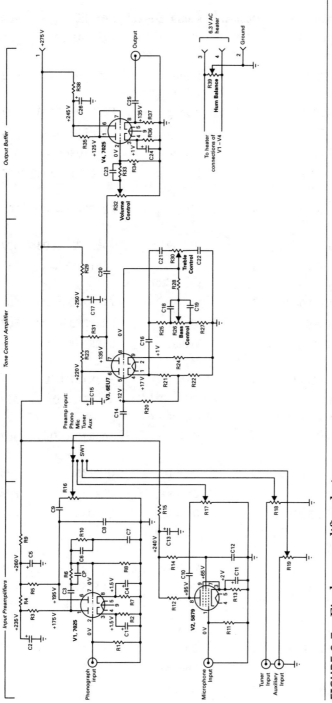

FIGURE 9.7   Final preamplifier design

TABLE 9.10 Parts List for the Final Preamplifier Design

Part	Description	Quantity	Manufacturer	Part No.	Allied Stock No.
C1, C4, C11	25 µF, 50 V, electrolytic	3	Vishay	TVA1306-E3	507-1089
C2, C5, C15, C17, C26	22 µF, 450 V, electrolytic	5	Illinois Capacitor	226TTA450M	613-0367
C3, C16	0.1 µF, 400 V	2	Vishay	225P10494XD3	507-0247
C6	0.0033 µF, 600 V	1	Vishay	715P33256JD3	507-0337
C7	0.01 µF, 400 V	1	Vishay	225P10394XD3	507-0642
C8	180 pF, 500 V, mica	1	Cornell-Dubilier	CD15FD181JO3F	862-0547
C9, C12, C20, C25	0.22 µF, 400 V	4	Vishay	225P22494YD3	507-0251
C10, C14	0.047 µF, 400 V	2	Vishay	715P47354LD3	507-0342
C13	47 µF, 450 V, electrolytic	1	Illinois Capacitor	476TTA450M	613-0369
C18, C22	0.0022 µF, 600 V	2	Vishay	715P22256JD3	507-0332
C19	0.022 µF, 400 V	1	Vishay	225P22394XD3	507-0643
C21	220 pF, 600 V, ceramic	1	Vishay	562R5GAT22	507-0815
C23	100 pF, 1000 V disc	1	Vishay	561R10TCCT10	507-0889
C24	47 µf, 50 v, electrolytic	1	Illinois Capacitor	476TTA050M	613-0342
CP	25 pF, 1000 V ceramic	1	Vishay	561R10TCCQ25	507-0725
R1[1]	47 kΩ, 0.5 W	1	Ohmite	OL4735E	296-4788
R2, R7	2.7 kΩ, 0.5 W	2	Ohmite	OL2725E	296-6170
R3, R5, R12, R25, R28, R31, R35, R37	100 kΩ, 0.5 W	8	Ohmite	OL1045E	296-4773
R4	39 kΩ, 0.5 W	1	Ohmite	OL3935E	296-4784
R6, R14, R20	470 kΩ, 0.5 W	3	Ohmite	OL4745E	296-4789
R8	680 kΩ, 0.5 W	1	Tyco	LR1F680K	437-0431
R9, R29, R38	15 kΩ, 2 W	3	Ohmite	OY153KE	296-2333
R10, R15, R23	22 kΩ, 0.5 W	3	Ohmite	OL2235E	296-4779
R11	2.2 mΩ, 0.5 W	1	Ohmite	OL2255E	296-6466
R13, R24, R36	1 kΩ, 0.5 W	3	Ohmite	OL1025E	296-4772

(continued)

TABLE 9.10 Parts List for the Final Preamplifier Design (*Continued*)

Part	Description	Quantity	Manufacturer	Part No.	Allied Stock No.
R16, R17, R18, R19	1 mΩ potentiometer, linear	4	Honeywell/ Clarostat	RV4LAYSA105A	753-2670
R21	1.5 kΩ, 0.5 W	1	Ohmite	OL1525E	296-4774
R22	15 kΩ, 0.5 W	1	Ohmite	OL1535E	296-4775
R26, R30, R32	1 mΩ potentiometer, audio taper	3	Precision Electronic Components[2]	KA1051S28	
R27	10 kΩ, 0.5 W	1	Ohmite	OL1035E	296-1483
R33, R34	220 kΩ, 0.5 W	2	Ohmite	OL2245E	296-4780
R39	100 Ω, 1 W, potentiometer[3]	1	Honeywell/ Clarostat	381L100	753-8428
SW1	Rotary switch	1	Electroswitch	C4D0206N-A	747-6690
	Four-terminal barrier strip	1	Molex	38720-6204	607-0071
V1, V4	7025/12AX7 tube	2	Electro-Harmonix, Genalex, others		Tube Depot[3]
V2	5879 tube	1	NOS		Tube Depot
V3	6EU7 tube	1	Sovtek, others		Tube Depot
	¾-inch #4 threaded standoff	8	Keystone Electronics	1895	839-0781
	9-pin tube socket	4	Belton	P-ST9-601	AES[4]

[1] Optimum value depends on the type of magnetic pickup used. Follow manufacturer's recommendations.
[2] This device can be ordered from Digi-Key Electronics (www.digikey.com).
[3] Vacuum tubes are available from a number of suppliers, including Tube Depot (www.tubedepot.com), The Tube Store (http://thetubestore.com), Tube World (www.tubeworld.com), and other suppliers.
[4] Available from Antique Electronic Supply (www.tubesandmore.com/).

A parts list for the preamplifier is given in Table 9.10.

For a stereo preamplifier, the parts listed in Table 9.10 should be doubled. The reader may wish to use ganged potentiometers for the bass, treble, and volume controls, rather than separate controls for each channel. For dual controls, the Precision Electronic Components #KKA1051S28-ND may be used.

The construction considerations outlined previously apply here as well. The preliminary test readings documented in Tables 9.6 and 9.7 should also be used for this circuit.

PWB Design

For the final preamplifier design, a single-board PWB implementation was developed that incorporates all components, except for the power supply. The overall component layout is shown in Figure 9.8. For this design, a ground plane was used on the component side of the board to reduce hum and stray noise. The vacuum tubes are mounted on the chassis and connected with short jumpers to the PWB. The wire connection codes are given in Table 9.11.

The PWB is connected to the chassis through the mounting holes for the vacuum tubes. The tube sockets are conventional chassis-mounted devices, with jumper leads extending from the active pins to the PWB. The sockets are held away from the PWB using standoffs. This approach permitted the sockets to be firmly mounted on the

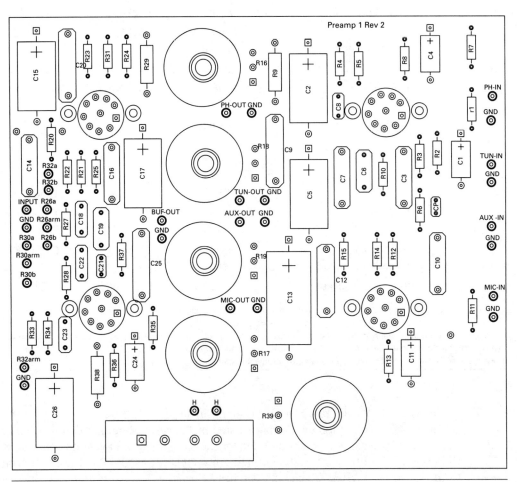

FIGURE 9.8 Component layout for the preamplifier PWB

TABLE 9.11   PWB Solder-On Connection Wiring Codes

PWB Code	Function	Connection Color Code	Component
PH-IN	Phonograph input	Yellow	
GND	Ground	Black	
TUN-IN	Tuner input	Yellow	
GND	Ground	Black	
AUX-IN	Auxiliary input	Yellow	
GND	Ground	Black	
MIC-IN	Microphone input	Yellow	
GND	Ground	Black	
PH-OUT	Phono preamp output to SW1	Shielded	
GND	Ground	Shield	
TUN-OUT	Tuner to SW1	Shielded	
GND	Ground	Shield	
AUX-OUT	Auxiliary output to SW1	Shielded	
GND	Ground	Shield	
MIC-OUT	Microphone preamp output to SW1	Shielded	
GND	Ground	Shield	
H	6.3 V AC heater	Green	
H	6.3 V AC heater	Green	
INPUT	Tone control stage input	Yellow	
GND	Ground	Black	
R26a	High terminal of R26	Yellow	R26
R26arm	R26 wiper	Yellow	
R26b	Low terminal of R26	Yellow	
R30a	High terminal of R30	Yellow	R30
R30arm	R30 wiper	Yellow	
R30b	Low terminal of R30	Yellow	
R32a	High terminal of R32	Yellow	R32
R32arm	R32 wiper	Yellow	
R32b	Low terminal of R32	Yellow	
GND	Ground	Black	
BUF-OUT	Output of buffer amplifier	Shielded	
GND	Ground	Shield	

chassis, rather than being physically supported by the PWB. This mounting method also takes advantage of the shielding capability of the steel chassis.

Heater connections to the vacuum tubes are made directly on the sockets using interconnecting wire tightly twisted together to minimize hum.

When making adjustments on the potentiometers mounted on the PWB, use an insulated screwdriver. While the potentiometer shaft is grounded, other components on the board operate at B+ voltages. Use extreme caution.

 It is not possible to reproduce the PWB circuit traces for the preamplifier board in this book. The layout file is, however, available from the author at the VacuumTubeAudio.info website, as detailed at the end of the book.

Chapter 10

Project 3: 15 W Audio Power Amplifier

Project 3 is a high-fidelity 15 W audio power amplifier. The design objectives are as follows:

- Audio output = 15 W
- Total harmonic distortion = 1 percent or less
- Intermodulation distortion = 1.5 percent or less
- Frequency response within ±1 dB, 20 Hz to 20 kHz
- Total hum and noise = 75 dB below rated output (input terminals shorted)
- Total project cost for parts below $750
- Hand-wired terminal strip construction

As shown in Figure 10.1, the amplifier utilizes a high-gain pentode input stage followed by a triode-based phase inverter to drive the push-pull output tubes. A rectifier feeds the choke-input filter, which supplies the needed operating voltages to the tubes.

Circuit Description

A high-gain pentode voltage amplifier is used as the input stage for the audio power amplifier [1]. The output of this stage is direct-coupled to the control grid of a triode split-load type of phase inverter. The use of direct coupling between these stages minimizes phase shift and, consequently, increases the amount of inverse feedback that may be used without danger of low-frequency instability.

A low-noise 7199 tube, which contains a high-gain pentode section and a medium-μ triode section in one envelope, fulfills the active-component requirement for both the pentode input stage and the triode phase inverter. Potentiometer R1 in the input circuit of the 7199 pentode section functions as the volume control for the amplifier.

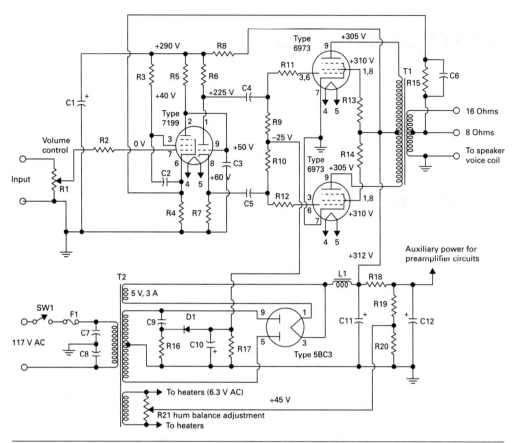

FIGURE 10.1 Schematic diagram of a 15 W audio power amplifier. Typical operating voltages are given for reference. (*After* [1].)

The plate and cathode outputs of the phase inverter, which are equal in amplitude and opposite in phase, are used to drive a pair of pentode 6973 beam power tubes used in a class AB$_1$ push-pull output stage. The 6973 output tubes are biased for class AB$_1$ operation by the fixed negative voltage applied to the control grid circuit from the rectifier circuit. Fixed bias is used because a class AB amplifier provides highest efficiency and least distortion for this bias method.

Transformer T1 couples the audio amplifier output to the speaker. The taps on the secondary of the transformer match the plate-to-plate impedance of the output stage to the voice coil impedance of an 8 Ω or 16 Ω loudspeaker. Negative feedback of 19.5 dB is coupled from the secondary of the output transformer to the cathode of the input stage to reduce distortion and improve circuit stability.

The transformer-coupled AC input power is converted to DC operating power for the amplifier stages by the 5BC3 full-wave rectifier. Heater power for the amplifier tubes and the rectifier is obtained from the 6.3 V and 5 V secondary windings, respectively, on power transformer T2.

Fixed bias operation of the output stage requires that the power supply provide good voltage regulation, as the plate current of the 6973 tubes vary considerably with signal level. The conventional choke-input type of power supply used provides the required stability. The fixed bias for the output stage is obtained from one-half of the high-voltage secondary winding of power transformer T2 through a capacitance-resistance voltage divider and a silicon rectifier. Potentiometer R21 connected across the 6.3 V secondary winding of transformer T2 provides a hum balance adjustment for the amplifier. The wiper arm of the potentiometer is connected to the junction of a resistive voltage divider across the output of the power supply. The resulting positive bias voltage applied to the tube heaters minimizes heater-to-cathode leakage and substantially reduces hum.

If the amplifier oscillates, or "motorboats," reverse the ground and feedback connections at the secondary of output transformer T1. Damage to the output tubes and/or other devices in the final stage can result from sustained operation in a motorboat condition.

Parts List

Table 10.1 lists the electronic components needed to build this amplifier, along with manufacturer part numbers and stock numbers from one of the major parts houses.[1] These parts are, of course, available from a number of manufactures and suppliers. The details given here are for the convenience of the reader. Considerations relating to chassis components and assembly techniques are addressed in other chapters.

For a stereo amplifier, the parts quantities listed in the table should be doubled. For stereo operation from a single power supply, power transformer T2 must be replaced by one that has a higher current rating. A transformer rated for 375-0-375 V rms at 200 mA (or greater) is recommended, such as the Hammond 274BX.

At the time of this writing, the 6973 was being manufactured by one or more tube companies. It appeared that the 7199 and 5BC3 were available only from new old stock (NOS). One alternative to using the 5BC3 is to substitute silicon rectifiers. A pair of NTE-575[2] rectifiers should suffice.

It is recommended that a matched pair be used for the 6973 output tubes. A matched pair is more expensive than purchasing two randomly selected devices. However, because this amplifier does not include provisions for balancing the output tubes through adjustment of cathode current, the target distortion levels may not be possible without a matched pair.

---

[1] Allied Electronics, 7151 Jack Newell Blvd. S., Fort Worth, Texas, 76118, USA, https://www.alliedelec.com.
[2] Allied stock number 935-0064.

TABLE 10.1   Parts List for the Initial Build of the 15 W Audio Amplifier

Part	Description	Quantity	Manufacturer	Part No.	Allied Stock No.
C1	47 µF, 450 V electrolytic	1	Illinois Capacitor	476TTA450M	613-0369
C2, C4, C5	0.22 µF, 400 V	3	Vishay	715P22454MD3	507-0664
C3	3.3 pF, 1000 V, ceramic or mica	1	Vishay	561R10TCCV33	507-1284
C6	120 pF, 1000 V, ceramic or mica	1	Vishay	562R10TST12	507-0731
C7, C8	0.047 µF, 400 V	2	Vishay	715P47354LD3	507-0342
C9	0.022 µF, 600 V	1	Vishay	715P22356KD3	507-0334
C10	100 µF, 50 V electrolytic	1	Illinois Capacitor	107TTA050M	613-0343
C11	100 µF, 450 V, electrolytic	1	Illinois Capacitor	107TTA450M	613-0836
R1	1 megohm potentiometer	1	Honeywell/ Clarostat	RV4NAYSD105A	753-1320
R2	10 kΩ, 0.5 W	1	Ohmite	OL1035E	296-1483
R3	820 kΩ, 0.5 W	1	Vishay	VR37000008203JR500	648-0216
R4	825 Ω, 0.5 W	1	Vishay	RN65D8250F	895-0098
R5	220 kΩ, 0.5 W	1	Ohmite	OL2245E	296-4780
R6, R7	15 kΩ, 2 W	2	Ohmite	OY153KE	296-2333
R8	3.9 kΩ, 2 W	1	Ohmite	OY392K	296-2362
R9, R10	100 kΩ, 0.5 W	2	Ohmite	OL1045E	296-4773
R11, R12	1 kΩ, 0.5 W	2	Ohmite	OL1025E	296-4772
R13, R14	100 Ω, 2 W	2	Ohmite	OY101KE	296-2318
R15	8.2 kΩ, 1 W	1	Ohmite	OX822KE	296-2309
R16	15 kΩ, 1 W	1	Ohmite	OM1535E	296-4796
R17	68 kΩ, 0.5 W	1	RCD Components	CF50S-683-JTW	840-0233
R18	4.7 kΩ, 2 W	1	Ohmite	ON4725E	296-4823
R19	270 kΩ, 1 W	1	Ohmite	OM2745E	296-4803

(continued)

TABLE 10.1   Parts List for the Initial Build of the 15 W Audio Amplifier (*Continued*)

Part	Description	Quantity	Manufacturer	Part No.	Allied Stock No.
R20	47 kΩ, 0.5 W	1	Ohmite	OL4735E	296-4788
R21	100 Ω, 2W potentiometer	1	Honeywell/ Clarostat	RV4NAYSD101A	753-1301
D1	Rectifier, 1 A, 600 V,	1	NTE	116	935-0298
L1	Choke: 5 H 200 mA DC resistance 65 Ω	1	Hammond[1]	193H	Newark
T1	Output transformer: Primary– 6600 Ω plate-to-plate 20 W Secondary– 4, 8, 16 Ω	1	Hammond	1620	Newark
T2	Power transformer: Primary– 120 V Secondary– 360-0-360 V rms, 138 mA 6.3 V, 5 A 5 V, 3 A	1	Hammond	274AX	Newark
V1	7199 tube	1	JAN-Philips, GE, others		Tube Depot[2]
V2, V3	6973 tube	2	Electro-Harmonix, others		Tube Depot
V4	5BC3 tube	1	Various		Tube Store[2]

[1] Hammond Manufacturing Company, Inc., 475 Cayuga Rd., Cheektowaga, NY, 14225, USA: www.hammondmfg.com. Available from Newark Electronics (www.newark.com).

[2] Vacuum tubes are available from a number of suppliers, including Tube Depot (www.tubedepot.com), The Tube Store (http://thetubestore.com), Tube World (www.tubeworld.com), and Antique Electronic Supply (www.tubesandmore.com).

Construction Considerations

The 15 W amplifier described in this chapter can be constructed in any number of ways. In this example, this unit was hand-wired using terminal strips on an aluminum chassis. A component layout diagram is not provided for this project because the chassis used is relatively large and, therefore, the parts density of the circuit is rather low. The reader may wish to sketch out a layout plan before proceeding.

A parts list for the chassis components is given in Table 10.2. Bulk supplies needed for project assembly are listed in Chapter 6 (Table 6.6).

The construction of the amplifier is conventional in approach (see Figure 10.2). As noted in Chapter 8, this circuit was built on a test bed. Construction techniques are essentially identical for either the single-purpose chassis or the test bed.

TABLE 10.2 Chassis Parts List for the Initial Build Circuit

Description	Quantity	Manufacturer	Part No.	Allied Stock No.	Notes
Chassis, 17-in × 10-in × 3-in	1	Hammond	1444-32	806-3042	
9-pin tube socket	3	Belton	SK-B-VT9-ST-2		Tube Depot
9-pin novar socket	1		SK-9PINX		Tube Depot
AC line fuse holder	1	Bussmann	HTB-32I-R	740-0652	
Fuse, 3 A, slo-blow	1	Bussmann	MDL-3-R	740-0748	
AC power cord	1	Alpha Wire	615 BK078	663-7020	6 ft. power cord
Power switch	1	Eaton	7504K4	757-4201	
Knob	1	Davies	1110	543-1110	Volume control
Chassis feet	4	Bud	F-7264-A	736-7264	
RCA connector	1	Neutrix	NF2D-B-0	514-0004	Audio connector
6-term barrier strip	1	Waldom Molex	38720-3206	607-0066	Speaker output
Terminal strip	8	Cinch	54A	750-6628	

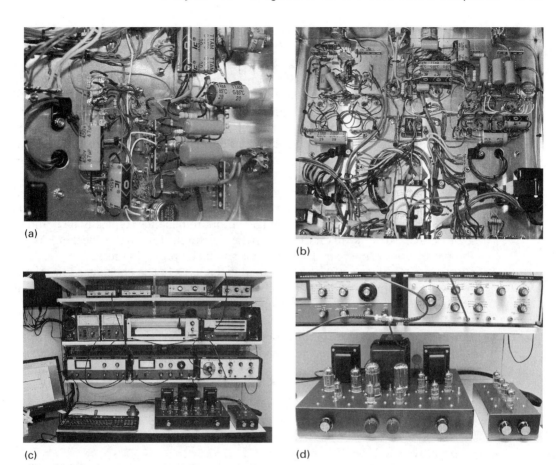

(a)

(b)

(c)

(d)

FIGURE 10.2   Completed 15 W amplifier: (*a*) chassis bottom view of 15 W amplifier circuit, (*b*) chassis bottom view of overall test bed, (*c*) unit under test on the bench, (*d*) close-up of power amplifier test bed (left) and preamp test bed (right).

With regard to components, the only differences between the test bed implementation of the 15 W amplifier and the parts specified in Tables 10.1 and 10.2 are the transformers and the aluminum chassis. Specifically:

- **Chassis**   A larger chassis was used,[3] making it possible to build both the 15 W amplifier described in this chapter and the 30 W amplifier described in Chapter 11 on the same piece of aluminum. For readers planning on building just the 15 W amplifier, the chassis specified in Table 10.2 should be quite adequate.

---

[3] Chassis size = 17-in × 14-in × 3-in.

- **Output transformer** A Hammond #1620 is specified in Table 10.1. The transformer used in the test bed was a Hammond 1650H. The two are identical except that the 1620 is rated for 20 W and the 1650H is rated for 40 W output. Note that the Hammond 1620 and 1650H output transformers have a specified method of connecting the secondary windings to properly interface with 4, 8, or 16 Ω loads. Carefully follow the appropriate wiring diagram provided with the device (or available for download from the manufacturer's website). Note also that the feedback circuit of R16/C6 is connected to the 8 Ω output, as documented in Figure 10.1. The chassis parts list in Table 10.2 specifies a 6-terminal barrier strip for the speaker connections. The assumption here is that the secondary windings of the 1650H will be brought out to the terminal strip, and the necessary jumpers installed to provide the desired output impedance (plus ground). Readers may wish to simplify this arrangement and strap the secondary winding for a single output impedance (typically 8 Ω) and use a 2-terminal barrier strip on the back panel for connection to the speaker.[4] As with the power transformer, any unused leads should be properly dressed.
- **Choke** A Hammond #193H is specified in Table 10.1. The choke used in the test bed was a Hammond 193LP. The 193H is rated for 5 H, 200 mA, and DC resistance of 65 Ω. The 193LP is a high-performance potted device rated for 5 H, 300 mA, and DC resistance of 55 Ω.
- **Power transformer** A Hammond #274AX is specified in Table 10.1. The transformer used in the test bed was a Hammond 300BX. The 274AX is rated as follows:

primary = 120/240 V
secondary = 360-0-360 V rms @ 138 mA, 6.3 V @ 5 A, 5 V @ 3 A

The 300BX is rated as follows:

primary = 120/240 V
secondary = 400-0-400 V rms @ 288 mA, 5 V @ 1.2 A, 5 V @ 3 A, 6.3 V @ 6 A

It is worth noting that with the 300BX transformer used in the test bed, the screen voltage for the 6973 output tube is marginally above the rated maximum design value (see Chapter 7). For a finished product, the 274AX transformer (or equivalent) should be used.

The considerations outlined in Chapter 8 regarding power transformer connections apply here as well. One caution is worth repeating, however.

It is important to note that the 5BC3A rectifier tube heater draws a rated current of 3 A. For that reason, the proper 5 V secondary output winding on the power transformer must be used. There are two available 5 V output windings on the Hammond 300BX: one rated for 1.2 A and the other rated for 3 A.

For experimenters who want to build a show-quality amplifier, it is advisable to follow the test bed approach used by the author. This permits experimentation in circuit design and layout, and helps make the final unit look and perform as expected. For projects such as those described in this book, the major expenses are concentrated

---

[4] Such as the Molex 38720-3202; Allied stock number 607-0063.

in transformers and—to a lesser extent—the vacuum tubes. These high-cost items can be repurposed for the final, show-quality unit.

Initial Checkout

After the amplifier has been assembled, the following steps are recommended to confirm proper construction:

1. Remove AC power from the amplifier (unplug the AC line cord).
2. Remove all tubes.
3. Using an ohmmeter, confirm that the resistance values listed in Table 10.3 are observed on the tube socket connections. All readings are with reference to ground. Readings within ±5 percent are considered normal.
4. If any deviations from the expected values listed in Table 10.3 are observed, recheck wiring and components for correct installation. Do not advance to the next step until the resistance measurements are within tolerance.

TABLE 10.3 Typical Ohmmeter Readings with Power Removed and Tubes Removed

Tube	Pin No.	Typical Reading	Notes
V1, 7199	1	~340 kΩ	Varies due to filter capacitors in circuit
	2	~545 kΩ	Varies due to filter capacitors in circuit
	3	~1.1 MΩ	Varies due to filter capacitors in circuit
	4	~40 kΩ	Heater
	5	~40 kΩ	Heater
	6	820 Ω	
	7	10 kΩ	With volume control fully ccw (1 MΩ fully cw)
	8	15 kΩ	
	9	~545 kΩ	Varies due to filter capacitors in circuit
V2, 6973	1	~325 kΩ	Varies due to filter capacitors in circuit
	2	–	
	3	~170 kΩ	Varies due to filter capacitors in circuit
	4	~40 kΩ	Heater
	5	~40 kΩ	Heater
	6	~170 kΩ	Varies due to filter capacitors in circuit
	7	0 Ω	
	8	~325 kΩ	Varies due to filter capacitors in circuit
	9	~325 kΩ	Varies due to filter capacitors in circuit

(*continued*)

TABLE 10.3   Typical Ohmmeter Readings with Power Removed and Tubes Removed (*Continued*)

Tube	Pin No.	Typical Reading	Notes
V3, 6973	1	~375 kΩ	Varies due to filter capacitors in circuit
	2	–	
	3	~170 kΩ	Varies due to filter capacitors in circuit
	4	~40 kΩ	Heater
	5	~40 kΩ	Heater
	6	~170 kΩ	
	7	0 Ω	
	8	~1 MΩ	Varies due to filter capacitors in circuit
	9	~200 kΩ	Varies due to filter capacitors in circuit
V4, 5BC3	1	~325 kΩ	Varies due to filter capacitors in circuit
	2	–	
	3	~325 kΩ	Varies due to filter capacitors in circuit
	4	–	
	5	~45 Ω	
	6	–	
	7	–	
	8	–	
	9	~45 Ω	

5. After the resistance checks have been successfully completed, install V4 (the 5BC3 rectifier tube) and apply power to the unit. Be very careful, as high voltages are present in the circuit.

6. Using extreme caution, check for proper voltages at the tube connection points listed in Table 10.4. All voltages are measured with respect to ground using a high-impedance voltmeter. Variations within ±10 percent are typical. These measurements can be taken from the top of the chassis by probing into the socket connection points.

TABLE 10.4   Typical Voltmeter Readings with V1, V2, and V3 Removed and Power Applied

Tube	Pin No.	Typical Reading	Notes
The following readings are with a B+ supply voltage of 390 V DC. See note 1.			
V1, 7199	1	386 V	Note 2
	2	376 V	Note 2
	3	355 V	Note 2

(continued)

TABLE 10.4   Typical Voltmeter Readings with V1, V2, and V3 Removed and Power Applied (*Continued*)

Tube	Pin No.	Typical Reading	Notes
V1, 7199	4	57 V	Heater, varies with setting of hum balance control
	5	57 V	Heater, varies with setting of hum balance control
	6	0 V	
	7	0 V	
	8	0 V	
	9	375 V	Note 2
V2, 6973	1	384 V	Note 2
	2	–	
	3	−36 V	
	4	57 V	Heater, varies with setting of hum balance control
	5	57 V	Heater, varies with setting of hum balance control
	6	−36 V	
	7	0 V	
	8	384 V	Note 2
	9	384 V	Note 2
V3, 6073	1	384 V	Note 2
	2	–	
	3	−36 V	
	4	57 V	Heater, varies with setting of hum balance control
	5	57 V	Heater, varies with setting of hum balance control
	6	−36 V	
	7	0 V	
	8	384 V	Note 2
	9	384 V	Note 2

Notes:

1 The B+ power supply output with no load is approximately 390 V DC for the transformer described in the test bed implementation (Hammond #300BX, plate winding = 800 V ct.). The transformer specified in the parts list in Table 10.1 (Hammond #274AX, plate winding = 720 V ct.) will provide a slightly lower B+ supply voltage in the range of 350 V DC no load. Note that these voltages will decrease when the tubes are installed and the supply is loaded.

2 If the Hammond #274A is used for T2, the voltage measurement should be scaled appropriately; that is, readings of approximately 40 V lower would be typical for B+ voltages.

7. If variations in the expected voltage readings are found, remove all power from the unit (unplug the AC line cord and confirm that the power supply has discharged to 0 V) and recheck components and wiring. Do not advance to the next step unless the voltages check correctly.

8. Remove AC power from the unit.

9. Install all tubes.

10. Set the volume control to the minimum (counterclockwise) position.

11. Connect a loudspeaker to the output terminals.

12. Connect the AC line cord and switch the power on. Observe the tubes for any signs of overheating or other unexpected behavior. The filaments should be lit. Remove power if any problems are detected or if noticeable hum can be heard from the speaker. Recheck wiring and components as needed.

13. If oscillation or motorboating is observed, immediately remove power from the unit. After the filter capacitors have fully discharged, switch the feedback circuit leads, as described previously in this chapter. It is important that an oscilloscope be connected to the output terminals, in addition to a speaker and perhaps other test equipment, when the amplifier is first powered up. It is possible that high-frequency oscillation can occur that cannot be reproduced by the loudspeaker or heard by the builder. Monitoring with an oscilloscope is the only certain way of determining whether any high-frequency oscillations are occurring (or very low-frequency oscillations for that matter). When connecting test equipment to the output terminals, remember that one side of the output transformer is connected to ground. Incorrect placement of test leads can short-circuit the output.

14. Reapply power and confirm that the motorboating and/or oscillation has ceased. If problems continue, disconnect power and recheck the installation of all components, particularly the input stage (V1).

15. After these preliminary tests have been successfully completed, measurements should be taken on the amplifier as detailed in the next section.

16. The amplifier should be operated only with its rated load applied. Operation into an open-circuit or short-circuit load can result in damage to components.

Measured Performance

Upon completion of the amplifier, measurements were taken at output levels of 1 W, 5 W, 10 W, 12.5 W, and 15 W (the maximum rated output) into an 8 Ω load. The following test equipment was used:

- Heathkit IG-18 audio generator
- Heathkit IG-1275 lin/log sweep generator
- Heathkit IM05258 harmonic distortion analyzer
- Heathkit SM-5248 intermodulation distortion analyzer

- Heathkit IM-5238 AC voltmeter
- Heathkit IM-4100 frequency counter
- Heathkit IO-4510 oscilloscope

For the measurements, an 8 Ω 20 W noninductive resistor was connected to the 8Ω output terminals. The test equipment was connected to the same 8Ω output. The measured performance of the 15 W audio amplifier is given in Table 10.5.

TABLE 10.5   Measured Performance of 15 W Audio Amplifier

Conditions	Parameter	Measured Value	Notes
Measurements at 1 W power output (1 kHz into 8 Ω)	Input level	0.31 V rms	Gain control fully clockwise
	Frequency response	±1 dB, 6 Hz to 60 kHz	Note 1
	THD	0.3%	Note 2
	IMD	0.59%	4:1 mix ratio
	Noise, unweighted (shorted input terminals)	−71 dB	Gain control fully clockwise
Measurements at 5 W power output (1 kHz into 8 Ω)	Input level	0.78 V rms	Gain control fully clockwise
	Frequency response	±1 dB, 9 Hz to 60 kHz	Note 3
	THD	1.0%	Note 2, note 4
	IMD	1.2%	4:1 mix ratio
	Noise, unweighted (shorted input terminals)	−79 dB	Gain control fully clockwise
Measurements at 10 W power output (1 kHz into 8 Ω)	Input level	0.93 V rms	Gain control fully clockwise
	Frequency response	±1 dB, 10 Hz to 60 kHz	Note 5
	THD	1.2%	Note 2, note 6
	IMD	1.2%	4:1 mix ratio
	Noise, unweighted (shorted input terminals)	−81 dB	Gain control fully clockwise
Measurements at 12.5 W power output (1 kHz into 8 Ω)	Input level	1.1 V rms	Gain control fully clockwise
	Frequency response	±1 dB, 20 Hz to 50 kHz	Note 7
	THD	1.5%	Note 2, note 8
	IMD	1.4%	4:1 mix ratio
	Noise, unweighted (shorted input terminals)	−81 dB	Gain control fully clockwise

(continued)

TABLE 10.5 Measured Performance of 15 W Audio Amplifier (*Continued*)

Conditions	Parameter	Measured Value	Notes
Maximum operating point (note 9)	Input level	1.2 V rms	Gain control fully clockwise
	Output level	11 V rms	With 8 Ω load
	Computed output power	15.125 W	
	THD at 1 kHz	0.71%	
	Noise, unweighted (shorted input terminals)	−82 dB	Gain control fully clockwise

Notes:
1 The frequency response of this amplifier is quite impressive. At 1 W output, the 3 dB points were 5 Hz and 144 kHz. Curiously, there is a bump of +4 dB at approximately 90 kHz.
2 THD measurements were taken at 20 Hz, 100 Hz, 1 kHz, 10 kHz, and 20 kHz unless otherwise noted. The worst (highest) measurement is the one quoted.
3 At 5 W output, the 3 dB points were 6 Hz and 95 kHz.
4 At 5 W output, THD was below 0.6% from 23 Hz to 18 kHz.
5 At 10 W output, the 3 dB points were 7 Hz and 83 kHz.
6 At 10 W output, THD was well below 1% from 23 Hz to 18 kHz.
7 At 12.5 W output, the 3 dB points were 9 Hz and 70 kHz.
8 At 12.5 W output, THD was below 1% from 25 Hz to 15 kHz.
9 The clipping point was identified visually on an oscilloscope with a 1 kHz input signal; it was observed to be symmetrical. The input level was then reduced just enough to eliminate any clipping. This was deemed to be the maximum operating point of the amplifier.

The following observations and comments can be made following construction and performance measurements on this amplifier:

- The 7199 medium-μ sharp-cutoff pentode input/phase-splitter tube is no longer being manufactured, and as a result the device can be difficult to find (and expensive). The author used this tube in the 15 W circuit because of the favorable comments the NOS device (particularly the JAN/Philips version) has received in certain quarters. The 30 W amplifier described in Chapter 11 calls for the 7199 as well, but for that implementation, a replacement device, the 6U8A, was used. The 6U8A is billed as a functional replacement, although the pinout is different. Keep in mind that the 7199 and 6U8A are not pin-compatible. As documented in Chapter 11, certain performance issues were observed with the 6U8A in the initial build of the 30 W circuit. The author did not try the 6U8A in the 15 W circuit, although he knows of no reason it should not provide acceptable results.
- As noted previously, this amplifier was built on a test bed with the basic power supply described in Chapter 8. The operating voltages shown in Figure 10.1 apply to the Hammond 274AX transformer shown in the parts lists (Table 10.1), not those of the higher-voltage Hammond 300BX. The nominal difference in plate/screen operating voltages is approximately 40 V. Figure 10.3 repeats the schematic given in Figure 10.1, but with the voltages typical for a 350 V B+ supply. Note that the bias voltage at the junction of R9 and R10 is −35 V DC. The original RCA design called for −25 V, but with the higher plate voltage this was found to be too low a value, causing excessive idling current through the 6973 output tubes.

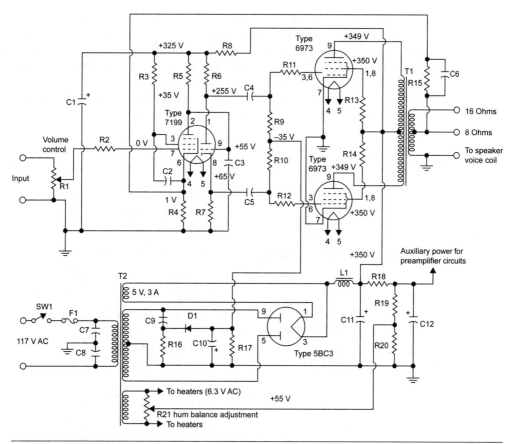

FIGURE 10.3   Typical operating voltages using the Hammond 300BX power transformer

- The original RCA design in [1] describes this as a 15 W amplifier. Strictly speaking, it certainly is. The amplifier will make slightly more than 15 W into an 8 Ω load with low distortion at mid-band audio frequencies (0.71 percent at 1 kHz). However, the distortion performance of the amplifier at full output at very low and very high audio frequencies is rather high (more than 2 percent). The distortion numbers at the extremes could be improved by lowering the grid bias on the output tubes, but that has the drawback of increasing the idling current. A different pair of output tubes might have made a difference, but that was not explored in this case. In any event, the author would prefer to bill this circuit as a 12.5 W amplifier, since at that level good performance can be achieved. As a practical matter, it probably doesn't make much difference, since nearly all listening in a home environment is done in the 1 to 5 W average power output level, depending on speaker efficiency (and the presence of neighbors).
- The 6973 beam power tube can be operated in an alternate configuration wherein the screen grids are connected to taps on the output transformer primary, rather than through small-value resistors connected to the B+ supply. Not all output

transformers have provisions for such connectors. Fortunately, the Hammond 1620 does; taps are provided for the screen grid of each tube in a push-pull arrangement at 40 percent of the plate winding. Hammond (and perhaps other manufacturers) describes this as "ultra-linear operation." With billing like that, the author had to try it out. The results were impressive, as documented in Table 10.6, which compares the test results obtained with the original RCA design and the results obtained with the screen grid tap configuration.

TABLE 10.6   Measured Performance of 15 W Audio Amplifier with Screen Tap Option

Conditions	Parameter	Value With Original Design	Value With Screen Tap Option
Measurements at 1 W power output	Input level	No substantive difference	
	Frequency response	No substantive difference	
	THD	0.3%	0.2%
	IMD	0.59%	0.23%
	Noise, unweighted (shorted input terminals)	No substantive difference	
Measurements at 5 W power output	Input level	No substantive difference	
	Frequency response	±1 dB, 9 Hz to 70 kHz	±1 dB, 8 Hz to 80 kHz
	THD	1.0%	0.3%
	IMD	1.2%	0.28%
	Noise, unweighted (shorted input terminals)	No substantive difference	
Measurements at 10 W power output	Input level	No substantive difference	
	Frequency response	No substantive difference	
	THD	1.2%	0.37%
	IMD	1.2%	0.29%
	Noise, unweighted (shorted input terminals)	No substantive difference	
Measurements at 12.5 W power output	Input level	No substantive difference	
	Frequency response	±1 dB, 20 Hz to 50 kHz	±1 dB, 12 Hz to 50 kHz
	THD	1.5%	0.8% (note 1)
	IMD	1.4%	0.98%
	Noise, unweighted (shorted input terminals)	No substantive difference	

Note:
1 THD at 20 Hz measured 0.8%. Within the passband of 25 Hz to 20 kHz, THD measured 0.4% or less.

As documented in Table 10.6, the improvement in distortion was significant with the screen tap option. The peak power available from the amplifier was slightly lower but close to the original design. The overall gain and noise floor were essentially unchanged.

The bias setting for the screen tap configuration was set as follows:

- With no input applied, the bias supply was set for approximately –35V, as specified in the tube data sheet for this mode of operation.
- The input signal was set to provide peak output from the amplifier, as described in Table 10.5 Note 9 (just below clipping).
- With an input signal of 20 kHz, distortion was measured and the bias control was adjusted for minimum reading. It was observed that power output did not change significantly within the bias range of –20V to –35 V DC.
- The input signal was changed to 20 Hz, and distortion was measured. The bias control was adjusted again for minimum distortion. The optimal setting was observed to roughly coincide for each frequency extreme. In the event that the optimal settings for 20 Hz and 20 kHz do not coincide, split the difference between the two settings.

For this particular implementation, the lowest distortion point was found to be at a bias of about –25 V DC. While varying the bias, it was observed that distortion on either side of this point increased, albeit only slightly. Be careful when adjusting bias, since lower settings (less negative) will increase the idling current. For test purposes, setting the bias for best performance is reasonable. For long-term operation, however, it is advisable to find the ideal operating point and then back off somewhat; for example, if –25 V is found to be the optimal setting, reduce bias to –30 V for long-term operation. Experience and performance should be the guide here. Keep in mind that the bias may need to be adjusted over time as the tubes age; when replacing tubes, repeat the bias setting adjustment.

Note that it may not be possible to achieve the desired low-frequency distortion target, even with careful adjustment of the bias. The output transformer used in the test bed (Hammond 1620) is rated for operation down to 30 Hz. The characteristics at 20 Hz are not documented, and as such, it may not be advisable to force the amplifier (through low bias) to make the target numbers. Instead, in this case, it may be better to aim for 30 Hz as the low-end distortion measurement. This issue is more critical at high power levels than at low levels. As documented in Table 10.6, at 12.5 W the THD measured at 20 Hz was 0.8 percent; that reading dropped to 0.4 percent at 25 Hz. In this case, pushing the tubes to get the last measure of performance below 25 Hz is inadvisable.

With this circuit configured for screen tap operation, the difference in distortion performance with bias set at the optimal level and distortion with bias set to a higher (more negative) level is quite small, as documented in Table 10.7. It can be seen that the increase in distortion is quite modest, and generally an acceptable tradeoff for reduced power consumption and longer tube life. Excessive idling current can result in short operational life of the output tubes and/or tube failure.

TABLE 10.7 Measured Distortion Performance of 15 W Audio Amplifier with Different Bias Points

Conditions	Parameter	Measurement With Optimal Bias (−25 V)	Measurement With Higher Bias (−33 V)
Measurements at 1 W power output	THD	0.2%	0.3%
	IMD	0.23%	0.7%
Measurements at 5 W power output	THD	0.3%	0.8%
	IMD	0.28%	1.6%
Measurements at 10 W power output	THD	0.37%	1.0%
	IMD	0.29%	1.6%
Measurements at 12.5 W power output	THD	0.8%	1%, 25 Hz to 20 kHz note 1
	IMD	0.98%	1.7%

Note:
1 At 12.5 W output, THD at 20 Hz measured 2.1%.

Final Power Amplifier Design

Using the basic RCA power amplifier shown in Figure 10.1 and the experience gained in building and testing the circuit, the author arrived at a final 12.5 W power amplifier design, as shown in Figure 10.4. Few substantive changes have been made, except to use the output transformer screen tap option described previously and to replace with 5BC3 tube rectifier with silicon diodes. The 5BC3 can be difficult to find (even at specialty suppliers), and so a solid-state solution was deemed appropriate.

Additional modifications to the basic power supply include adding an AC line filter and surge-limiting resistor in the primary circuit, along with the relay cutoff described in Chapter 8. Adjustment of the bias level has also been provided through the addition of potentiometer R17. An adjustment range of about −20 V to −35 V is typical. Zener diodes have been added to the bias circuit and hum balance circuit to prevent excessive voltages on the grids of the output tubes and heaters, respectively.

A parts list for the final 12.5 W amplifier design is given in Table 10.8. For stereo operation, the quantities shown should be doubled—with the exception of the power supply components. For the choke-input power supply filter, a bank of two 210 μF, 400 V capacitors is used to provide approximately 105 μF for the filter. The capacitors are bypassed with 100 kΩ, 2 W resistors to equalize the voltages across the capacitors and provide a discharge path when power is removed. A B+ voltage output is provided at terminal #1 for a preamplifier, if used.

Note that a linear taper potentiometer is specified for R1 (the volume control). A linear taper was chosen over an audio taper under the assumption that the power amplifier will be fed by a preamplifier with an integrated volume control. As such, the

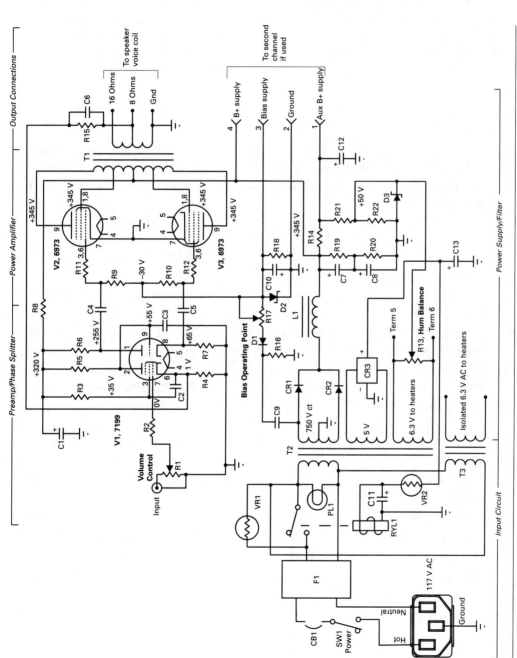

FIGURE 10.4   Final design of 12.5 W power amplifier

function of R1 is only to set the nominal input level of the amplifier and would not typically be adjusted during use.

The power transformer and choke listed in Table 10.8 assume that the supply will be used to drive two channels.

The output transformer specified in Table 10.8—Hammond 1620A—is a newer version of the Hammond 1620 specified in Table 10.1. The two devices have similar specifications, but the 1620A secondary is simpler to configure. With the 1620, the secondary is configured for operation at 4, 8, or 16 Ω by changing winding connections. The 1620A is more straightforward and simpler to use.

TABLE 10.8   Parts List for the Final 12.5 W Audio Amplifier Design

Part	Description	Quantity	Manufacturer	Part No.	Allied Stock No.
C1, C12	47 µF, 450 V electrolytic	2	Illinois Capacitor	476TTA450M	613-0369
C2, C4, C5	0.22 µF, 600 V	3	Vishay	715P22456MD3	507-0335
C3	3.3 pF, 1000 V, ceramic or mica	1	Vishay	561R10TCCV33	507-1284
C6	120 pF, 1000 V, ceramic or mica	1	Vishay	562R10TST12	507-0731
C7, C8	210 µF, 400 V	2	Cornell-Dubilier	CGS211T450R4C	857-0260
	Mounting bracket for C7, C8	2	Cornell-Dubilier	VR3A	852-7532
C9	0.022 µF, 600 V	1	Vishay	715P22356KD3	507-0334
C10, C11	100 µF, 100 V electrolytic	2	Illinois Capacitor	107TTA100M	613-0354
C13	1,000 µF, 25 V, electrolytic	1	Illinois Capacitor	108TTA025M	613-0327
R1	1 megohm potentiometer	1	Honeywell/Clarostat	RV4NAYSD105A	753-1320
R2	10 kΩ, 0.5 W	1	Ohmite	OL1035E	296-1483
R3	820 kΩ, 0.5 W	1	Vishay	VR37000008203JR500	648-0216
R4	825 Ω, 0.5 W	1	Vishay	RN65D8250F	895-0098
R5	220 kΩ, 0.5 W	1	Ohmite	OL2245E	296-4780
R6, R7	15 kΩ, 2 W	2	Ohmite	OY153KE	296-2333
R8	3.9 kΩ, 2 W	1	Ohmite	OY392K	296-2362
R9, R10	100 kΩ, 0.5 W	2	Ohmite	OL1045E	296-4773

(continued)

TABLE 10.8   Parts List for the Final 12.5 W Audio Amplifier Design (*Continued*)

Part	Description	Quantity	Manufacturer	Part No.	Allied Stock No.
R11, R12	1 kΩ, 0.5 W	2	Ohmite	OL1025E	296-4772
R13	100 Ω, 2W potentiometer	1	Honeywell/Clarostat	RV4LAYSA101A	753-1001
R14	4.7 kΩ, 2 W	1	Ohmite	OY472KE	296-2369
R15	8.2 kΩ, 1 W	1	Ohmite	OX822KE	296-2309
R16	15 kΩ, 1 W	1	Ohmite	OM1535E	296-4796
R17	25 k, 2 W, potentiometer	1	Honeywell/Clarostat	53C225K	753-1206
R18	68 kΩ, 2 W	1	Ohmite	OY683KE	296-2384
R19, R20	100 kΩ, 2 W	2	Ohmite	OY104KE	296-2322
R21	270 kΩ, 2 W	1	Ohmite	OY274KE	296-7117
R22	47 kΩ, 2 W	1	Ohmite	OY473KE	296-7121
D1	Rectifier, 1 A, 600 V	1	NTE	116	935-0298
D2, D3	Zener, 82 V, 5 W	2	ON Semiconductor	1N5375BG	568-0596
CR1, CR2	1A, 1000 V DC rectifier		NTE	NTE575	935-0064
CR3	Bridge rectifier, 100 V, 2 A	1	NTE	NTE166	935-6256
VR1	Power thermistor, 2.5 Ω cold	1	GE	CL-30	837-0030
VR2	Thermistor, 20 Ω cold	1	Honeywell	ICL1220002-01	254-0094
F1	AC line filter	1	Qualtek	848-10/003	689-3564
SW1	SPST switch	1	Eaton	7504K4	757-4201
CB1	Circuit breaker, 8 A	1	Tyco	W58-XB1A4A-8	886-8876
RYL1	5 V DC relay, contacts 240 V AC, 10 A	1	Magnecraft	9AS1D52-5	850-0123
PL1	117 V AC pilot lamp	1	Dialight	249-7841-1431-574	511-0276
L1	Choke: 5 H 200 mA DC resistance 65 Ω	1	Hammond[1]	193H	Newark

(*continued*)

TABLE 10.8 Parts List for the Final 12.5 W Audio Amplifier Design (*Continued*)

Part	Description	Quantity	Manufacturer	Part No.	Allied Stock No.
T1	Output transformer:	1	Hammond	1620A	Newark
	Primary– 6600 Ω plate-to-plate 20 W				
	Secondary– 4, 8, 16 Ω				
T2	Power transformer:	1	Hammond	274BX	Newark
	Primary– 120 V				
	Secondary– 375-0-375 V rms, 201 mA 6.3 V, 5 A 5 V, 3 A				
T3	Filament transformer:	1	Hammond	167Q6	Newark
	Primary– 120 V				
	Secondary– 6.3 V, 6 A				
V1	7199 tube	1	JAN-Philips, GE, others		Tube Depot[2]
V2, V3	6973 tube	2	Electro-Harmonix, others		Tube Depot
	9-pin tube socket	3	Belton	P-ST9-601	AES[3]
	3/4-inch threaded #4 standoff	10	Keystone Electronics	1895	839-0781
	4-terminal barrier strip	2	Molex	38720-6204	607-0071

[1] Hammond Manufacturing Company, Inc., 475 Cayuga Rd., Cheektowaga, NY, 14225, USA: www.hammondmfg.com. Available from Newark Electronics (www.newark.com).
[2] Vacuum tubes are available from a number of suppliers, including Tube Depot (www.tubedepot.com), The Tube Store (http://thetubestore.com), Tube World (www.tubeworld.com), and Antique Electronic Supply (www.tubesandmore.com).
[3] Antique Electronic Supply (www.tubesandmore.com).

After assembly, the same procedures described previously in this chapter should be used to confirm proper construction. In particular:

- See Table 10.3 for typical ohmmeter readings of the circuit with power and tubes removed. Ignore the entries for the 5BC3, since that device is not used in this application.
- See Table 10.4 for typical voltmeter readings with tubes removed and power applied.

Be certain to follow all of the steps outlined in the section "Initial Checkout."

PWB Design

For the final 12.5 W power amplifier design, a single-board PWB implementation was developed that incorporates all components, except for the transformers and filter capacitors C7/C8. The overall component layout is shown in Figure 10.5. For this design, a ground plane was used on the component side of the board to reduce hum and noise. The wire connection codes are given in Table 10.9.

The PWB is connected to the chassis through the mounting holes for the vacuum tubes. Additional supporting posts are accommodated. The tube sockets are conventional chassis-mounted devices, with jumper leads extending from the active pins to the PWB. The sockets are held away from the PWB using standoffs. This approach permitted the sockets to be firmly mounted on the chassis, rather than being physically supported by the PWB. This approach also keeps heat above the chassis and away from the PWB, which can degrade over time due to excessive heat.

The mounting method chosen for the vacuum tubes permits easy modification of the input circuit to accommodate use of a 6U8A tube for V1. As mentioned previously, the 6U8A is a functional replacement for the 7199. The pinout, however, is different. The changes needed to use the 6U8A are detailed in Table 10.10. The 6U8A is more readily available, and generally less expensive than the 7199. Keep in mind that the PWB was designed for the 7199; to use a 6U8A, the socket-to-PWB connections need to be wired as detailed in Table 10.10.

For readers making the change to a 6U8A, be certain to check the resistance measurements documented in Table 10.3 and Table 10.4 with the appropriate pin number changes.

Heater connections to the vacuum tubes are made directly on the sockets using interconnecting wire tightly twisted together to minimize hum. The approach taken here is to keep heater voltages off the PWB. The heater wires should be dressed against the chassis.

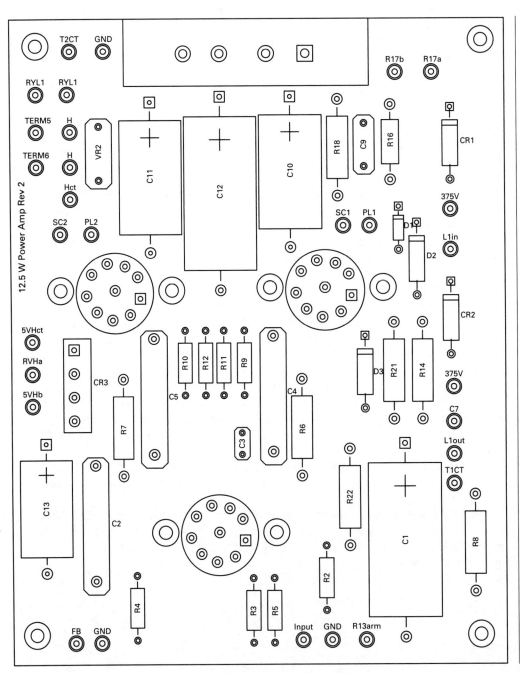

FIGURE 10.5   Component layout for the preamplifier PWB

TABLE 10.9   PWB Solder-On Connection Wiring Codes. (Note that the color codes for the transformers listed in this table may vary. Consult the device manufacturer before proceeding.)

PWB Code	Function	Connection Color Code	Component
PL1	Output transformer plate 1 connection	Blue	T1
SC1	Output transformer screen 1 connection	Blue/yellow	
PL2	Output transformer plate 2 connection	Brown	
SC2	Output transformer screen 2 connection	Brown/yellow	
T1CT	Output transformer center tap	Red	
INPUT	Input to amplifier	Yellow	
GND	Ground	Black	
FB	Feedback from output transformer	Shielded	
GND	Ground	Shield	
375V	Plate winding	Red	T2
375V	Plate winding	Red	
T2CT	Plate winding center tap	Red/yellow	
H	6.3 V heater	Green	
H	6.3 V heater	Green	
Hct	6.3 V heater center tap	Green/yellow	
5Vha	5 V heater	Yellow	
5VHb	5 V heater	Yellow	
5VHct	5V heater center tap	Yellow/black	
L1in	Choke in	Black	L1
L1out	Choke out	Black	
C7	Capacitor filter bank (C7/C8, R19/R20)	Red	C7/R19
GND	Filter bank ground	Black	
R13arm	Wiper arm of R13	Green	R13
R17a	R17 high terminal	White	R17
R17b	R17 low terminal	White	
RYL1	Relay 1 coil	White	RYL1
RYL1	Relay 1 coil	White	

TABLE 10.10 7199 to 6U8A Socket-to-PWB Wiring Modifications

Socket Pin No. for Original Design 7199	Socket Pin No. for 6U8A	Socket/PWB Wiring Change
2, pentode plate	6, pentode plate	Pin 6 on socket to pin 2 on PWB
6, pentode cathode	7, pentode cathode	Pin 7 on socket to pin 6 on PWB
7, pentode grid #1	2, pentode grid #1	Pin 2 on socket to pin 7 on PWB

Note It is not possible to reproduce the PWB circuit traces for the 12.5 W amplifier in this book. The layout file is, however, available from the author at the VacuumTubeAudio.info website, as detailed at the end of the book.

Readers, of course, may also build this amplifier using hand-wired terminal strips. The considerations given previously for construction practices apply here as well.

Chapter 11

Project 4: 30 W Audio Power Amplifier

Project 4 is a high-fidelity 30 W audio power amplifier. The design objectives of the amplifier are as follows:

- Audio output = 30 W
- Total harmonic distortion = 1 percent or less
- Intermodulation distortion = 1.5 percent or less
- Frequency response within ±1 dB, 20 Hz to 20 kHz
- Total hum and noise = 80 dB below rated output (input terminals shorted)
- Total project cost for parts below $950
- Hand-wired terminal strip construction

As shown in Figure 11.1, the amplifier utilizes a high-gain pentode input stage followed by a triode-based phase inverter to drive the push-pull output tubes. A rectifier feeds the capacitor-input filter, which supplies the needed operating voltages to the tube circuits. This amplifier shares the same basic architecture of the 15 W audio power amplifier described in Chapter 10.

Circuit Description

A high-gain pentode voltage amplifier is used as the input stage for the audio power amplifier [1]. The output of this stage is direct-coupled to the control grid of a triode split-load type phase inverter. The use of direct coupling between these stages minimizes phase shift and, consequently, increases the amount of inverse feedback that may be used without danger of low-frequency instability.

A low-noise 7199 tube, which contains a high-gain pentode section and a medium-μ triode section in one envelope, fulfills the active-component requirement

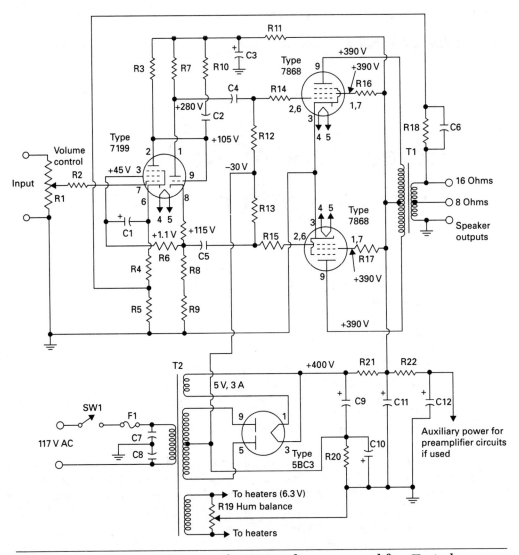

FIGURE 11.1   Schematic diagram of a 30 W audio power amplifier. Typical operating voltages are shown for reference. (*After* [1].)

for both the pentode input stage and the triode phase inverter. Potentiometer R1 in the input circuit of the 7199 pentode section functions as the volume control for the amplifier.

The plate and cathode outputs of the phase inverter, which are equal in amplitude and opposite in phase, are used to drive a pair of cathode-connected 7868 beam power tubes used in a class AB$_1$ push-pull output stage. The 7868 output tubes are biased for class AB$_1$ operation by the fixed negative voltage applied to the control grid circuit from the rectifier circuit. Fixed bias is used because a class AB amplifier provides highest efficiency and least distortion for this bias method.

Transformer T1 couples the audio amplifier output to the speaker. The taps on the secondary of the transformer match the plate-to-plate impedance of the output stage to the voice coil impedance of an 8 Ω or 16 Ω loudspeaker. Negative feedback of 19.5 dB is coupled from the secondary of the output transformer to the cathode of the input stage to reduce distortion and improve circuit stability.

The transformer-coupled AC input power is converted to DC operating power for the amplifier stages by the 5BC3 full-wave rectifier. Heater power for the amplifier tubes and the rectifier are obtained from the 6.3 V and 5 V secondary windings, respectively, on power transformer T2.

Fixed bias operation of the output stage requires that the power supply provide good voltage regulation, as the plate current of the 7868 tubes vary considerably with the signal level. The fixed bias is obtained from a resistive network in the negative leg of the power supply. Potentiometer R19 connected across the 6.3 V secondary winding of transformer T2 provides a hum balance adjustment for the amplifier.

If the amplifier oscillates or "motorboats," reverse the ground and feedback connections in the secondary of output transformer T1.

Parts List

Table 11.1 lists the electronic components needed to build this amplifier, along with manufacturer part numbers and stock numbers from one of the major parts houses.[1] Considerations relating to chassis components and assembly techniques are addressed in other chapters.

For a stereo amplifier, the parts quantities listed in the table should be doubled. For stereo operation from a single power supply, power transformer T2 must be replaced by one that has a higher current rating. A transformer rated for 400-0-400 V rms at 275 mA (or greater) is recommended, such as the Hammond 300BX.[2]

At the time of this writing, the 7868 tube was being manufactured by one or more tube companies. It appeared that the 7199 and 5BC3 were available only from new old stock (NOS). One alternative to using the 5BC3 is to substitute silicon rectifiers. A pair of NTE-575[3] rectifiers should suffice. It is recommended that a matched pair be used for the 7868 output tubes.

---

[1] Allied Electronics, 7151 Jack Newell Blvd. S., Fort Worth, Texas, 76118, USA. www.alliedelec.com. ·
[2] This is the device used in the test bed.
[3] Allied stock number 935-0064.

TABLE 11.1   Parts List for the Initial Build of the 30 W Audio Amplifier

Part	Description	Quantity	Manufacturer	Part No.	Allied Stock No.
C1	25 µF, 50 V, electrolytic	1	Vishay	TVA1306-E3	507-1089
C2	22 pF, 600 V, ceramic	1	Vishay	561R10TCCQ22	507-0880
C3	80 µF, 450 V, electrolytic	1	Vishay	TVA1716-E3	507-1123
C4, C5	0.22 µF, 600 V	2	Vishay	715P22456MD3	507-0335
C6	0.01 µF, 600 V	1	Vishay	715P10356KD3	507-0326
C7, C8	0.047 µF, 600 V	2	Vishay	715P47356LD3	507-0343
C9, C11	100 µF, 450 V, electrolytic[1]	4 4[1]	Illinois Capacitor Ohmite	107TTA450M OY104KE	613-0836 296-2322
C10	100 µF, 50 V, electrolytic	1	Illinois Capacitor	107TTA050M	613-0343
C12	22 µF, 450 V, electrolytic	1	Illinois Capacitor	226TTA450M	613-0367
R1	1 mΩ potentiometer	1	Honeywell/ Clarostat	RV4NAYSD105A	753-1320
R2	10 kΩ, 0.5 W	1	Ohmite	OL1035E	296-1483
R3	220 kΩ, 0.5 W	1	Ohmite	OL2245E	296-4780
R4	825 Ω, 0.5 W	1	Vishay	RN65D8250F	895-0098
R5	10 Ω, 0.5 W	1	Ohmite	OL1005E	296-4771
R6	182 kΩ, 0.5 W	1	Vishay	RN65D1823F	895-0287
R7, R8	15 kΩ, 2 W	2	Ohmite	OY153KE	296-2333
R9	1 kΩ, 2 W	1	Ohmite	OY102KE	296-2319
R10	22 kΩ, 0.5 W	1	Ohmite	OL2235E	296-4779
R11	2.2 kΩ, 2 W	1	Ohmite	OY222KE	296-2343
R12, R13	100 kΩ, 0.5 W	2	Ohmite	OL1045E	296-4773
R14, R15	1 kΩ, 0.5 W	2	Ohmite	OL1025E	296-4772
R16, R17	56 Ω, 0.5 W	2	Ohmite	OL5605E	296-6478
R18	270 Ω, 0.5 W	1	Ohmite	OL2715E	296-4782
R19	100 Ω, 2 W, potentiometer	1	Honeywell/ Clarostat	RV4NAYSD101A	753-1301

(continued)

TABLE 11.1   Parts List for the Initial Build of the 30 W Audio Amplifier (*Continued*)

Part	Description	Quantity	Manufacturer	Part No.	Allied Stock No.
R20	220 Ω, 10 W	1	Vishay	CW010220R0JE12	895-4733
R21	50 Ω, 10 W	1	Vishay	CW01050R00JE12	895-4145
R22	10 kΩ, 2W	1	Ohmite	ON1035E	296-1542
T1	Output transformer:	1	Hammond[2]	1650H	Newark
	Primary– 6600 Ω plate-to- plate 40 W				
	Secondary– 4, 8, 16 Ω				
T2	Power transformer:	1	Hammond	274BX	Newark
	Primary– 120 V				
	Secondary– 375-0-375 V rms, 201 mA 6.3 V, 6 A 5 V, 3 A				
V1	7199 tube	1	JAN-Philips, GE, others		Tube Depot[3]
V2, V3	7868 tube	2	Electro- Harmonix, others		Tube Depot
V4	5BC3 tube	1	Various		Tube Store

[1] The specified value for C9 and C11 is 50 µF at 500 V. This value of electrolytic capacitor is difficult to find. A simpler (and less expensive) approach is to use two 100 µF 450 V capacitors in series, with 100 kΩ 2 W resistors in parallel with each capacitor. This approach is reflected in the parts listed for C9 and C11.

[2] Hammond Manufacturing Company, Inc., 475 Cayuga Rd., Cheektowaga, NY,14225,USA: www.hammondmfg.com. Available from Newark Electronics (www.newark.com).

[3] Vacuum tubes are available from a number of suppliers, including Tube Depot (www.tubedepot.com), The Tube Store (http://thetubestore.com), Tube World (www.tubeworld.com), and Antique Electronic Supply (www.tubesandmore.com).

Construction Considerations

The 30 W amplifier described in this chapter can be constructed in any number of ways. For the purposes of this example, this unit was hand-wired using terminal strips on an aluminum chassis. A parts list for the chassis components is given in Table 11.2. Bulk supplies needed for project assembly are listed in Chapter 6 (Table 6.6).

TABLE 11.2 Chassis Parts List for the Initial Build Circuit

Description	Quantity	Manufacturer	Part No.	Allied Stock No.	Notes
Chassis, 17-in × 10-in × 3-in	1	Hammond	1444-32	806-3042	
9-pin tube socket	1	Belton	SK-B-VT9-ST-2		Tube Depot
9-pin novar socket	2	Belton	SK-9PINX		Tube Depot
AC line fuse holder	1	Bussmann	HTB-32I-R	740-0652	
Fuse, 3 A, slo-blow	1	Bussmann	MDL-3-R	740-0748	
AC power cord	1	Alpha Wire	615 BK078	663-7020	6 ft. power cord
Power switch	1	Eaton	7504K4	757-4201	
Knob	1	Davies	1110	543-1110	Volume control
Chassis feet	4	Bud	F-7264-A	736-7264	
RCA connector	1	Neutrix	NF2D-B-0	514-0004	Audio connector
6-term barrier strip	1	Waldom Molex	38720-3206	607-0066	Speaker output
Terminal strip	8	Cinch	54A	750-6628	

The construction of the amplifier is conventional in approach. As noted in Chapter 8, this circuit was built on a test bed. The Hammond 1650H output transformer has a specified method of connecting the secondary windings to properly interface with 4, 8, or 16 Ω loads. Carefully follow the appropriate wiring diagram provided with the device (or available for download from the manufacturer's website). As with the power transformer, any unused leads should be properly dressed.

The considerations outlined in Chapter 8 regarding power transformer connections apply here as well. One caution is worth repeating, however.

It is important to note that the 5BC3A rectifier tube heater draws a rated current of 3 A. For that reason, the proper 5 V secondary output winding on the power transformer must be used. There are two available 5 V output windings on the Hammond 300BX: one rated for 1.2 A and the other rated for 3 A. Note also that the filament leads for the 5BC3 carry the full B+ voltage. As such, hookup wire should be used that is rated for 600 V DC operation rather than the more common 300 V DC hookup wire. Dress the heater leads accordingly.

The chassis parts list in Table 11.2 specifies a 6-terminal barrier strip for the speaker connections. The assumption here is that the secondary windings of the 1650H will be brought out to the terminal strip, and the necessary jumpers installed to provide the desired output impedance (plus ground). Readers may wish to simplify this arrangement and strap the secondary winding for a single output impedance (typically 8 Ω) and use a 2-terminal barrier strip on the back panel for connection to the speaker.[4]

---

[4] Such as the Molex 38720-3202; Allied stock number 607-0063.

For experimenters who want to build a show-quality amplifier, it is advisable to follow the test bed approach used by the author. This permits experimentation in circuit design and layout, and helps make the final unit look and perform as expected. For projects such as those described in this book, the major expenses are concentrated in transformers and—to a lesser extent—the vacuum tubes. These high-cost items can be repurposed for the final, show-quality unit.

The completed unit is documented in Figure 11.2.

Initial Checkout

After the amplifier has been assembled, the following steps are recommended to confirm proper construction:

1. Remove AC power from the amplifier (unplug the AC line cord).
2. Remove all tubes.
3. Using an ohmmeter, confirm that the resistance values listed in Table 11.3 are observed on the tube socket connections. All readings are with reference to ground. Readings within ±5 percent are considered normal.

(a)

(b)

(c)

FIGURE 11.2   The 30 W audio amplifier: (*a*) bottom view, (*b*), detail view, (*c*) top view

TABLE 11.3 Typical Ohmmeter Readings with Power and Tubes Removed

Tube	Pin No.	Typical Reading	Notes
V1, 7199	1	<200 kΩ	Varies due to filter capacitors in circuit
	2	<200 kΩ	Varies due to filter capacitors in circuit
	3	~180 kΩ	Varies due to filter capacitor in circuit
	4	~50 Ω	Heater (hum balance control centered)
	5	~50 Ω	Heater (hum balance control centered)
	6	820 Ω	
	7	10 kΩ to 1 mΩ	Varies with setting of volume control
	8	16 kΩ	
	9	<200 kΩ	Varies due to filter capacitor in circuit
V2, 7868	1	–	Same as pin #7
	2	110 kΩ	
	3	0 Ω	
	4	~50 Ω	Heater (hum balance control centered)
	5	~50 Ω	Heater (hum balance control centered)
	6	–	Same as pin #2
	7	< 200 kΩ	Varies due to filter capacitor in circuit
	8	–	
	9	<200 kΩ	Varies due to filter capacitor in circuit
V3, 7868	1	–	Same as pin #7
	2	100 kΩ	
	3	0 Ω	
	4	~50 Ω	Heater (hum balance control centered)
	5	~50 Ω	Heater (hum balance control centered)
	6	–	Same as pin #2
	7	<200 kΩ	Varies due to filter capacitor in circuit
	8	–	
	9	<200 kΩ	Varies due to filter capacitor in circuit
V4, 5BC3	1	<200 kΩ	Varies due to filter capacitor in circuit
	2	–	
	3	<200 kΩ	Varies due to filter capacitor in circuit
	4	–	
	5	45 Ω	
	6	–	
	7	–	
	8	–	
	9	45 Ω	

4. If any deviations from the expected values listed in Table 11.3 are observed, recheck wiring and components for correct installation. Do not advance to the next step until the resistance measurements are within tolerance.
5. After the resistance checks have been successfully completed, install V4 (the rectifier tube) and apply power to the unit. Be very careful, since high voltages are present in the circuit.
6. Using extreme caution, check for proper voltages at the tube connection points listed in Table 11.4. All voltages are measured with respect to ground using a high-impedance voltmeter. Variations within ±10 percent are typical.

TABLE 11.4   Typical Voltmeter Readings with V1, V2, and V3 Removed and Power Applied

Tube	Pin No.	Typical Reading	Notes
The following readings are with a B+ supply voltage of 390 V DC.			
V1, 7199	1	390 V	
	2	385 V	
	3	0 V	
	4	–	Heater
	5	–	Heater
	6	0 V	
	7	0 V	
	8	0 V	
	9	385 V	
V2, 7868	1	–	
	2	–30 V	
	3	0 V	
	4	–	Heater
	5	–	Heater
	6	–	
	7	390 V	
	8	–	
	9	390 V	
V3, 7868	1	–	
	2	–30 V	
	3	0 V	
	4	–	Heater
	5	–	Heater
	6	–	
	7	390 V	
	8	–	
	9	390 V	

7. If variations in the expected voltage readings are found, remove all power from the unit (unplug the AC line cord and confirm that the power supply has completely discharged), and recheck components and wiring. Do not advance to the next step unless the voltages check correctly.

8. Remove AC power from the unit.

9. Install all tubes.

10. Set the volume control to the minimum (counterclockwise) position.

11. Set the hum balance control to the center of rotation.

12. Connect a loudspeaker to the output terminals.

13. Connect the AC line cord and switch the power on. Observe the tubes for any signs of overheating or other unexpected behavior. The filaments should be lit. Remove power if any problems are detected or if noticeable hum can be heard from the speaker. Recheck wiring and components as needed.

14. If oscillation or motorboating is observed, immediately remove power from the unit. After the filter capacitors have fully discharged, switch the feedback circuit leads, as described previously in this chapter. It is important that an oscilloscope be connected to the output terminals, in addition to a speaker and perhaps other test equipment, when the amplifier is first powered up. It is possible that high-frequency oscillation can occur that cannot be reproduced by the loudspeaker or heard by the builder. Monitoring with an oscilloscope is the only certain way of determining whether any high-frequency oscillations are occurring (or very low-frequency oscillations for that matter). When connecting test equipment to the output terminals, remember that one side of the output transformer is connected to ground. Incorrect placement of test leads can short-circuit the output.

15. Reapply power and confirm that the motorboating and/or oscillation has ceased. If problems continue, disconnect power and recheck the installation of all components, particularly the input stage (V1).

16. After these preliminary tests have been successfully completed, measurements should be taken on the amplifier as detailed in the next section.

17. The amplifier should be operated only with its rated load applied. Operation into an open-circuit or short-circuit load can result in damage to components.

Measured Performance

Upon completion of the amplifier, measurements were taken at output levels of 1 W, 5 W, 10 W, 20 W, and 25 W into an 8 Ω load. In addition, the clipping point of the amplifier was identified, again with an 8 Ω load. The following test equipment was used:

- Heathkit IG-18 audio generator
- Heathkit IG-1275 lin/log sweep generator
- Heathkit IM05258 harmonic distortion analyzer
- Heathkit SM-5248 intermodulation distortion analyzer
- Heathkit IM-5238 AC voltmeter
- Heathkit IM-4100 frequency counter
- Heathkit IO-4510 oscilloscope

For the measurements, an 8 Ω 30 W noninductive resistor was connected to the 8Ω output terminals. The test equipment was connected to the same 8Ω output.

Initial measurements on the amplifier as shown in Figure 11.1 yielded disappointing results. The frequency response was exceptional, and the noise floor was quite low. However, distortion, even at low power levels, was excessive. Table 11.5 shows the results of the initial tests.

Given the relatively poor distortion performance of the circuit at 5 W, no further measurements were taken on the amplifier in the original configuration.

Readers will recall the considerable improvement that the screen tap option had on distortion performance of the 15 W amplifier described in Chapter 10. Accordingly, the original RCA circuit shown in Figure 11.1 was modified to eliminate the screen grid resistors and instead connect the screen grids to the 40 percent taps provided on the Hammond 1650H output transformer. The improvement in performance was impressive, as documented in Table 11.6.

TABLE 11.5   Initial Measured Performance of the 30 W Audio Amplifier

Conditions	Parameter	Measured Value	Notes
Measurements at 1 W power output (1 kHz into 8 Ω)	Input level	0.12 V rms	Gain control fully clockwise
	Frequency response	±1 dB, 6 Hz to 50 kHz	Note 1
	THD	0.8%	Note 2
	IMD	0.8%	4:1 mix ratio
	Noise, unweighted (shorted input terminals)	−65 dB	Gain control fully clockwise
Measurements at 5 W power output (1 kHz into 8 Ω)	Input level	0.25 V rms	Gain control fully clockwise
	Frequency response	±1 dB, 7 Hz to 60 kHz	Note 3
	THD	2.8%	Note 2
	IMD	2.1%	4:1 mix ratio
	Noise, unweighted (shorted input terminals)	−72 dB	Gain control fully clockwise

Notes:

1 At 1 W output, the 3 dB points were 5 Hz and 70 kHz.

2 THD measurements were taken at 20 Hz, 100 Hz, 1 kHz, 10 kHz, and 20 kHz unless otherwise noted. The worst (highest) measurement is the one quoted.

3 At 5 W output, the 3 dB points were 6 Hz and 70 kHz.

TABLE 11.6   Measured Performance of the 30 W Audio Amplifier

Conditions	Parameter	Measured Value	Notes
Measurements at 1 W power output (1 kHz into 8 Ω)	Input level	0.12 V rms	Gain control fully clockwise
	Frequency response	±1 dB, 6 Hz to 60 kHz	Note 1
	THD	0.3%	Note 2
	IMD	0.12%	4:1 mix ratio
	Noise, unweighted (shorted input terminals)	−63 dB	Gain control fully clockwise
Measurements at 5 W power output (1 kHz into 8 Ω)	Input level	0.24 V rms	Gain control fully clockwise
	Frequency response	±1 dB, 6 Hz to 60 kHz	Note 3
	THD	0.3%	Note 2
	IMD	0.26%	4:1 mix ratio
	Noise, unweighted (shorted input terminals)	−68 dB	Gain control fully clockwise
Measurements at 10 W power output (1 kHz into 8 Ω)	Input level	0.35 V rms	Gain control fully clockwise
	Frequency response	±1 Hz, 9 Hz to 60 kHz	Note 4
	THD	0.6%	Note 2
	IMD	0.52%	4:1 mix ratio
	Noise, unweighted (shorted input terminals)	−71 dB	Gain control fully clockwise
Measurements at 20 W power output (1 kHz into 8 Ω)	Input level	0.46 V rms	Gain control fully clockwise
	Frequency response	±1 dB, 12 Hz to 50 kHz	Note 5
	THD	1.1%, 30 Hz to 20 kHz	Note 2, note 6
	IMD	0.69%	4:1 mix ratio
	Noise, unweighted (shorted input terminals)	−72 dB	Gain control fully clockwise
Measurements at 25 W power output (1 kHz into 8 Ω)	Input level	0.52 V rms	Gain control fully clockwise
	Frequency response	±1 dB, 14 Hz to 29 kHz	
	THD	1.1%, 30 Hz to 20 kHz	Note 2, note 7, note 8
	IMD	1.1%	4:1 mix ratio
	Noise, unweighted (shorted input terminals)	−75 dB	Gain control fully clockwise

(continued)

TABLE 11.6   Measured Performance of the 30 W Audio Amplifier (*Continued*)

Conditions	Parameter	Measured Value	Notes
Maximum operating point (note 9)	Input level	0.58 V rms	Gain control fully clockwise
	Output level	15 V rms	With 8Ω load
	Computed output power	28.125 W	
	THD at 1 kHz	0.75%	
	IMD	1.1%	4:1 mix ratio
	Noise, unweighted (shorted input terminals)	−76 dB	Gain control fully clockwise, note 10

Notes:
 1 The frequency response of the amplifier is impressive. At 1 W output, the 3 dB points were 5 Hz and 90 kHz.
 2 THD measurements were taken at 20 Hz, 100 Hz, 1 kHz, 10 kHz, and 20 kHz unless otherwise noted. The worst (highest) measurement is the one quoted.
 3 At 5 W output, the 3 dB points were 5 Hz and 90 kHz.
 4 At 10 W output, the 3 dB points were 6 Hz and 90 kHz.
 5 At 20 W output, the 3 dB points were 9 Hz and 70 kHz.
 6 At 20 W output, THD at 20 Hz was 1.8%.
 7 At 25 W output, the distortion at 20 Hz was quite high, in the neighborhood of 5%. At 25 Hz, distortion measured 1.6%. At 30 Hz, distortion had dropped to 1%.
 8 The bias voltages on the grids of the output tubes have a significant impact on distortion at the low and high end. It is recommended that the bias voltage be adjusted (if that capability exists) while monitoring THD at 20 kHz with 25 W output. The bias point should be adjusted for minimum THD. After making the adjustment, recheck distortion at all other frequencies. Note that in the test bed implementation, adjustable bias was included to allow for performance optimization. Note also that if the bias is set too low, excessive dissipation of the output tubes may result. Therefore, decrease bias only to the point that acceptable performance can be achieved. As noted in #7, distortion performance at 20 Hz with 25 W output was quite high. The Hammond 1650H output transformer is rated for operation down to 30 Hz. The performance at 20 Hz is not documented. Running down the bias on the output tubes to adjust for distortion improvement at 20 Hz in this situation is inadvisable.
 9 The clipping point was identified visually on an oscilloscope with a 1 kHz input signal; it was observed to be symmetrical. The input level was then reduced just enough to eliminate any clipping. This was deemed to be the maximum operating point of the amplifier. To calculate the maximum power output, the actual value of the load resistor is critical. It is important to remember that the value of the load resistor may change with operating temperature. In preliminary tests of this amplifier, the author measured a load resistor value when cold to be exactly 8.0 Ω. When hot, however, the resistance had dropped to 7.5 Ω.
10 The noise reading was improved by several dB through adjustment of the hum balance control.

Circuit Assessment

The following observations and comments can be made after construction of this amplifier. As mentioned previously, the circuit was built on the test bed, and therefore certain elements of the original RCA design shown in Figure 11.1 were not implemented, notably:

• The power supply filter used a 5 H choke in place of R21 since one was available on the test bed. The end result is probably the same, but given the choice of a choke or resistor, a choke is clearly the better option.

- The method of developing the necessary –30 V DC bias voltage for the output tubes seemed suboptimal, in that the center tap of the plate transformer is returned to ground through R20/C10. The author preferred another approach, made quite simple to implement by virtue of the 50 V "bias tap" on the Hammond 300BX power transformer used in the test bed. The bias tap winding provides about 50 V AC, referenced to the center tap of the plate winding. As such, it offers a convenient method of developing the necessary bias voltage for this circuit without placing the center tap above ground. In the test bed implementation, the 50 V tap was rectified and filtered, and then applied to a potentiometer, which allowed precise adjustment of bias. The author has no doubt the original RCA design shown in Figure 11.1 would work just fine. It was, however, not implemented since a more convenient method was available.
- In the original RCA design, a "hum balance" control was provided in the form of R19 that varied the reference to ground on the filaments. Readers will recall the 15 W amplifier documented in Chapter 10 used a positive DC voltage derived from the output of the power supply through a ladder divider to minimize hum. This seemed the better approach and was implemented in the test bed circuit. As noted in Table 11.6, adjustment of the hum balance control reduced the noise floor by several dB.

The following observations and comments apply to the test bed implementation of this circuit:

- The original RCA design called for a 7199 tube in the preamplifier/phase-splitter stage. This type is difficult to locate, and a 6U8A was identified by more than one source as a good replacement. In the original build of this circuit, the author used the 6U8A. It is important to note that the pinout of the 7199 is different from that of the 6U8A. While certain tube vendors offer the 6U8A as a "substitute" for the 7199 and provide the pin wiring changes needed to use the device, it is difficult to imagine making that type of modification on an amplifier after it has been built. After testing, the author went back to the 7199 for reasons detailed next.
- High-frequency oscillation at certain settings of the volume control (R1) was observed in the initial build of the circuit, mostly around the midpoint of travel. Installing the bottom plate on the chassis helped somewhat, as did shortening the component leads, notably capacitors C4 and C5. These devices, as specified in Table 11.1, are rated for operation at up to 600 V. Devices of this class are physically large and as such represent a challenge when it comes to controlling noise and preventing oscillation. The oscillation was observed to occur at approximately 60 kHz. After making the changes described here, the oscillation was reduced, but could not be eliminated completely. A replacement 6U8A was tried, but the oscillation remained at certain (center) settings of R1. The author changed the circuit to accommodate a 7199, and the oscillation disappeared. For this reason, the author cannot recommend use of the 6U8A in this particular circuit. Measurements documented in this chapter were made with the 7199.

- The reader will notice that the preamp/phase-splitter stage of the 30 W amplifier uses the same tube (7199) as the 15 W amplifier described in Chapter 10. There are some differences between the two circuits, in particular, the connections to grid #2. The author tried both configurations on the 30 W amplifier. No notable differences in performance were measured.
- The RCA circuit in [1] describes this as a 30 W amplifier. Well, after some effort the author was just not able to get that kind of power out of the circuit. In fairness to the venerable *RCA Receiving Tube Manual,* many of the key components identified in the original parts list are no longer available (or at least the author could not locate them). Notable among the component substitutions are the transformers. On the other hand, perhaps they just built them better "back in the day." Regardless, the author prefers to describe this as a 25 W amplifier since he was able to make that power on the bench with reasonable distortion. As a practical matter, this is probably a distinction without a difference, as very few people listen to sine waves at just below clipping. In any case, for most applications, 25 W output should be quite sufficient.

Expanding on the power output issue, while the amplifier failed to meet expectations in the original configuration (with 7868 screen grids connected to B+ through resistors R16 and R17), it exceeded expectations with the transformer screen tap option. For push-pull class AB_1 operation with the screens connected to B+ through low-value resistors, the data sheet [1] says the circuit should make 30 W with a plate voltage of 350 V (which was provided in the test bed). As noted earlier, the author could not get that much power out of the amplifier. On the other hand, the data sheet estimates power output at 23 W for class AB_1 operation with the screens connected to output transformer taps and with fixed bias. The author was able to get 25 W in that mode.

The following conclusions can be drawn:

- Use of the screen tap option reduces distortion but also reduces the maximum power output.
- The reduction in power output due to use of the screen tap was less than predicted "by the book."

Given the good distortion performance of the screen tap option, the tradeoff in slightly lower power output is a good bargain.

It is worth elaborating on the bias supply change described earlier. As noted, the test bed implementation provided for an adjustable bias voltage to the junction of R12 and R13 of –20 to –50 V. Prior to installation of the tubes, the bias control was set to provide –30 V as specified in the RCA design. During performance measurements, it was found that the crossover point of the output tubes with –30 V was not ideal, resulting in higher THD than was expected. To optimize the bias setting, a 1 kHz signal was delivered to the amplifier at a point sufficient to show flattening of the sine wave output on an oscilloscope. The input was then backed off just enough to eliminate any clipping. The bias voltage was then adjusted to provide the smoothest transition from positive to negative cycles as viewed on the scope. This point also coincided with the highest output from the amplifier. This setting of bias voltage was sufficient to provide good performance at low power levels. At higher levels, fine-tuning of the bias was found to improve (lower) distortion at low and high frequencies, as documented in Table 11.6, Note 8.

Readers should exercise caution when setting the bias. While a low setting may improve distortion performance, it will also increase the idling current through the output tubes. If excessively high, shortened tube life may result. It is recommended that the optimum bias point be identified and then backed off (bias made more negative) by some amount (e.g., 5 V) in order to extend tube life and reduce power consumption.

As shown in Figure 11.1, the original RCA design shows the feedback tap connected to the 16 Ω output, with a separate 8 Ω connection available from the output transformer. On the Hammond 1650H output transformer used in the test bed, the output impedance is configured using certain combinations of winding connections. The user selects 4, 8, or 16 Ω, but not a combination of those values. A newer version of the transformer, the Hammond 1650HA, does provide this flexibility. Because the preferred output for the amplifier was 8 Ω, the transformer secondary was configured accordingly. The measurements documented in Table 11.6 were taken with this arrangement.

Final Power Amplifier Design

Using the basic RCA power amplifier shown in Figure 11.1 and the experience gained in building and testing the circuit, the author arrived at a final power amplifier design, as shown in Figure 11.3. The following substantive changes have been made in the circuit:

- Screen taps were used on the output stage.
- The original power supply was replaced with the power supply shown in Chapter 8 (Figure 8.6), which provides for adjustable bias and other features.
- Wattage ratings for certain resistors were increased to provide for greater operating margin.
- The working voltage on the power supply filter capacitor was increased to provide for a greater operating margin.

Consideration was given to adding provisions to change the relative output of the push-pull output tubes through a low-value variable resistor (and associated components) to the cathode of V2 and V3. This change was not implemented, since adjustment of the bias provided for harmonic distortion readings that were quite low and well within the target numbers. As mentioned previously, the 7868 tubes used in the test bed implementation were matched devices. These tubes are readily available and reasonably priced. Therefore, no further changes to the amplifier were deemed necessary.

A parts list for the final 25 W amplifier design is given in Table 11.7. For stereo operation, the quantities shown should be doubled. Note that a linear taper

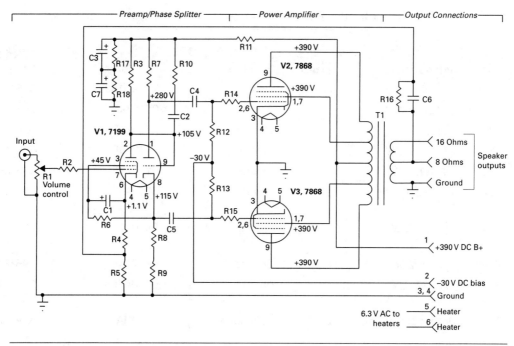

FIGURE 11.3   Final 25 W power amplifier design

TABLE 11.7   Parts List for the Final 30 W Audio Amplifier Design

Part	Description	Quantity	Manufacturer	Part No.	Allied Stock No.
C1	25 μF, 50 V, electrolytic	1	Vishay	TVA1306-E3	507-1089
C2	22 pF, 600 V, ceramic	1	Vishay	561R10TCCQ22	507-0880
C3, C7	160 μF, 450 V, electrolytic	2	Cornell-Dubilier	381LQ181M450K022	862-4752
C4, C5	0.22 μF, 600 V	2	Vishay	715P22456MD3	507-0335
C6	0.01 μF, 600 V	1	Vishay	715P10356KD3	507-0326
R1	1 mΩ potentiometer	1	Honeywell/Clarostat	RV4NAYSD105A	753-1320
R2	10 kΩ, 0.5 W	1	Ohmite	OL1035E	296-1483
R3	220 kΩ, 0.5 W	1	Ohmite	OL2245E	296-4780
R4	825 Ω, 0.5 W	1	Vishay	RN65D8250F	895-0098
R5	10 Ω, 0.5 W	1	Ohmite	OL1005E	296-4771

(continued)

TABLE 11.7　Parts List for the Final 30 W Audio Amplifier Design (*Continued*)

Part	Description	Quantity	Manufacturer	Part No.	Allied Stock No.
R6	182 kΩ, 0.5 W	1	Vishay	RN65D1823F	895-0287
R7, R8	15 kΩ, 2 W	2	Ohmite	OY153KE	296-2333
R9	1 kΩ, 2 W	1	Ohmite	OY102KE	296-2319
R10	22 kΩ, 0.5 W	1	Ohmite	OL2235E	296-4779
R11	2.2 kΩ, 2 W	1	Ohmite	OY222KE	296-2343
R12, R13	100 kΩ, 0.5 W	2	Ohmite	OL1045E	296-4773
R14, R15	1 kΩ, 0.5 W	2	Ohmite	OL1025E	296-4772
R16	270 Ω, 2 W	1	Ohmite	OY271KE	296-2348
R17, R18	100 kΩ, 2W	2	Ohmite	OY104KE	296-2322
T1	Output transformer: Primary– 6600 Ω plate-to-plate 40 W Secondary– 4, 8, 16 Ω	1	Hammond[1]	1650HA	Newark
V1	7199 tube	1	JAN-Philips, GE, others		Tube Depot[2]
V2, V3	7868 tube	2	Electro-Harmonix, others		Tube Depot
	9-pin tube socket	1	Belton	P-ST9-601	AES[3]
	9-pin novar socket	2	Belton	SK-9PINX	Tube Depot
	¾-inch #4 threaded standoff	10	Keystone Electronics	1895	839-0781
	4-terminal barrier strip	1	Molex	38720-6204	607-0071

[1] Hammond Manufacturing Company, Inc., 475 Cayuga Rd., Cheektowaga, NY,14225,USA: www.hammondmfg.com. Available from Newark Electronics (www.newark.com).
[2] Vacuum tubes are available from a number of suppliers, including Tube Depot (www.tubedepot.com), The Tube Store (http://thetubestore.com), Tube World (www.tubeworld.com), and Antique Electronic Supply (www.tubesandmore.com).
[3] Available from Antique Electronic Supply (www.tubesandmore.com/).

potentiometer is specified for R1 (the volume control). A linear taper was chosen over an audio taper under the assumption that the power amplifier will be fed by a preamplifier with an integrated volume control. As such, the function of R1 is only to set the nominal input level of the amplifier and would not typically be adjusted during use.

The output transformer specified in Table 11.7—Hammond 1650HA—is a newer version of the Hammond 1650H specified in Table 11.1. The two devices have similar specifications, but the 1650HA secondary is simpler to configure. As noted previously, with the 1650H, the secondary is configured for operation at 4, 8, or 16 Ω by changing winding connections. The 1650HA is more straightforward and simpler to use.

PWB Design

For the final 25 W power amplifier design, a single-board PWB implementation was developed that incorporates all components, except for the power supply. The overall component layout is shown in Figure 11.4. For this design, a ground plane was used on the component side of the board to reduce hum and noise. The wire connection codes are given in Table 11.8.

TABLE 11.8   PWB Solder-On Connection Wiring Codes. (Note that the color codes for the transformers listed in this table may vary. Consult the device manufacturer.)

PWB Code	Function	Connection Color Code	Component
PL1	Output transformer plate 1 connection	Blue	T1
SC1	Output transformer screen 1 connection	Blue/yellow	
PL2	Output transformer plate 2 connection	Brown	
SC2	Output transformer screen 2 connection	Brown/yellow	
CT	Output transformer center tap	Red	
GND	Ground	Black	
INPUT	Input to amplifier	Yellow	
GND	Ground	Black	
FB	Feedback from output transformer	Shielded	
GND	Ground	Shield	
H	6.3 V AC heater	Green	
H	6.3 V AC heater	Green	

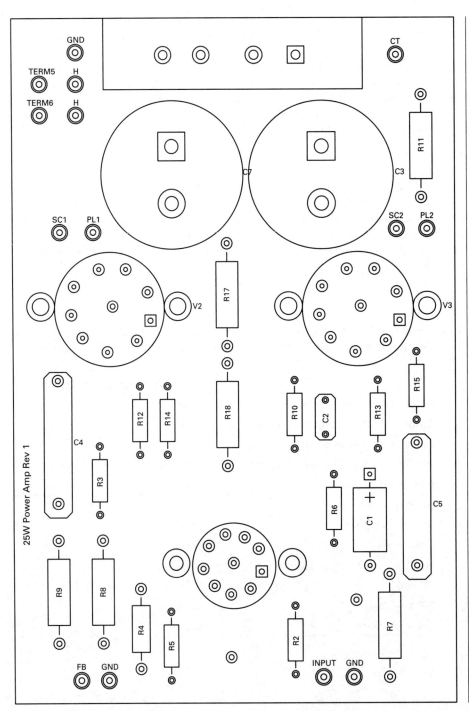

FIGURE 11.4 Component layout for the 25 W power amplifier PWB

The PWB is connected to the chassis through the mounting holes for the vacuum tubes. The tube sockets are conventional chassis-mounted devices, with jumper leads extending from the active pins to the PWB. The sockets are held away from the PWB using standoffs. This approach permitted the sockets to be firmly mounted on the chassis, rather than being physically supported by the PWB. This approach also keeps heat above the chassis and away from the PWB, which can degrade over time due to excessive heat.

Heater connections to the vacuum tubes are made directly on the sockets using interconnecting wire tightly twisted together to minimize hum. The approach taken here is to keep heater voltages off the PWB. The heater wires should be dressed against the chassis.

It is not possible to reproduce the PWB circuit traces for the 25 W amplifier in this book. The layout file is, however, available from the author at the VacuumTubeAudio.info website, as detailed at the end of the book.

Chapter 12

Putting It All Together

Having described a number of audio circuits in previous chapters, in this chapter we pull the various elements together to form finished consumer products. Three projects are described:

- Stereo preamplifier, based on the circuits and printed wiring board (PWB) design described in Chapter 9.
- 25 W stereo amplifier, based on the circuit and PWB design described in Chapter 10.
- 50 W stereo power amplifier, based on the circuit described in Chapter 11 and the power supply circuit described in Chapter 8.

The theory of operation, circuit design, and assembly details will not be repeated here. Instead, this chapter will focus on system assembly and related finishing techniques.

The Big Picture

When setting out to write this book, the author was thinking about the projects that could be included. The original idea was to have a 100 W stereo amplifier as the "big project." The number "100" sounds substantial—even impressive. Then reality set in. As a practical matter, it is difficult to build a vacuum tube stereo amplifier of more than about 50 W in a single chassis of reasonable size that will fit in someone's living room or study. The classic high-power amplifiers of yester-year were often single-channel "mono-block" units. For stereo operation, two are needed. Well, this takes up a lot of space and generates a lot of heat. And, did I mention the cost?

There are certainly applications for high-power vacuum tube amplifiers, but there is likely a much larger need for smaller, more compact units that can fit on a reasonably sized equipment shelf. With high power come large transformers, which are expensive and heavy. The heat generated by high-power amplifiers also deserves some consideration, particularly if long listening spans are envisioned.

So, that's the background. Attempting to pick the right niche for any product is difficult. Hopefully, this book will hit the mark for most readers.

Stereo Preamplifier

The preamplifier takes the final preamp design and PWB implementation described in Chapter 9 (Figure 9.7) to form a stereo preamp with four available inputs:

- Phonograph
- Microphone
- Auxiliary
- Tuner

An input switch selects the desired source, which is fed to a tone-control stage and finally to an output buffer amplifier. No power supply is included in the design, as the assumption is that this unit will be paired with a power amplifier. A multipin connector ties the power supply of the amplifier to the preamp chassis. This has the benefit of reducing cost and at the same time reducing noise from the supply that could be picked up by the low-level preamp circuits.

The completed preamp PWB is shown in Figure 12.1. Note the connection method for the tube sockets. The heater connections are made with twisted wire off the circuit board. This approach keeps AC voltages away from the preamp circuits and simplifies board layout. The back panel power connector pinout is given in Table 12.1.

Efficiently making the socket connections to the PWB requires a bit of practice. The recommended procedure is detailed in Figure 12.2. Be certain to wear protective eye glasses during this process, and at other steps in PWB assembly.

If a pin on a particular tube is not used, do not include a connecting wire. Take care to ensure that each connecting wire is mated with the proper pad on the PWB. Note also that no heater connections are made to the PWB, other than the 6.3 V AC source. Heater leads should be twisted and placed at approximately socket level. This will dress the leads against the chassis when the PWB is installed.

(a)

(b)

FIGURE 12.1   Preamp PWB: (*a*) component side, (*b*) foil side

TABLE 12.1   Preamp Power Connector Pinout

Pin No.	Function	Connects To	Notes
1	Ground	Terminal 2 of both preamplifier PWBs	
2	Pilot lamp	Front panel pilot lamp	117 V AC
3	Pilot lamp	Front panel pilot lamp	117 V AC
4	Auxiliary B+	Terminal 1 of both preamplifier PWBs	Approximately +275 V DC
5	Filament	Terminal 3 of both preamplifier PWBs	6.3 V AC for heaters
6	Filament	Terminal 4 of both preamplifier PWBs	6.3 V AC for heaters
7			Not used
8			Not used
9	Ground	Chassis ground	

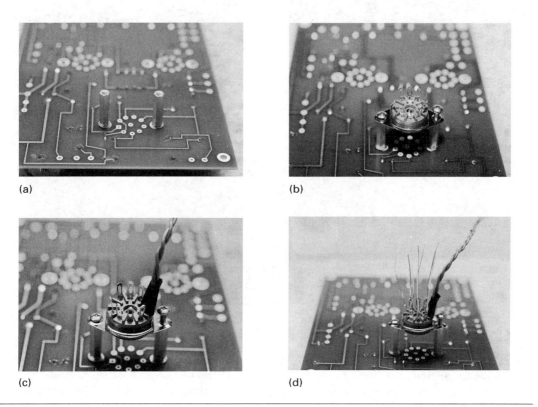

(a)   (b)   (c)   (d)

FIGURE 12.2   Wiring the tube sockets to the PWB: (*a*) mount the standoffs on the PWB; (*b*) loosely mount the socket upside down (with the proper orientation) on the standoffs; (*c*) make the heater connections; (*d*) attach connecting wires to the active socket pins, leaving about two inches of wire free; (*continued*)

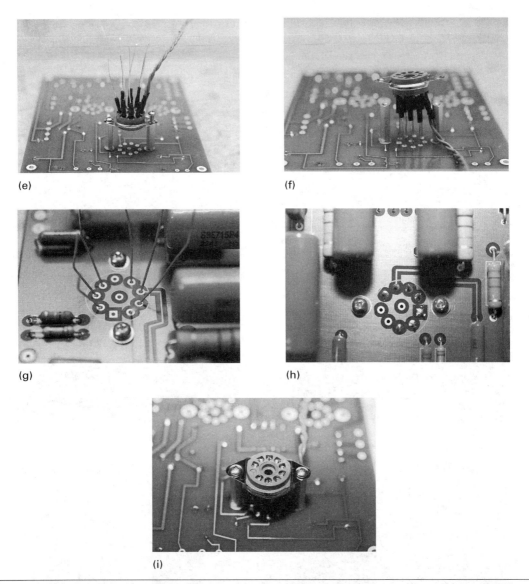

FIGURE 12.2 Wiring the tube sockets to the PWB: (*e*) place short sections of heat-shrink tubing over each wire/pin and process, then cut the wires in descending length around the socket to facilitate insertion in the PWB; (*f*) invert the socket and carefully match the connecting wires with their respective PWB pins; (*g*) pull the connecting wires tight toward the PWB, checking to make sure the face of the socket is parallel with the board; (*h*) solder from the component side and trim the excess wire, (*i*) finished installation

The screws holding the standoffs to the board and to the socket should be finger-tight during assembly. When the board is mated with the chassis, remove the screws on the sockets and fit the assembly up from the bottom of the chassis. Secure the sockets and PWB with screws from the top of the chassis. Once the screws have been threaded, tighten all component parts.

There is no mystery to populating a PWB. Having said that, some general guidelines are worth noting:

- Proceed in a logical manner, usually beginning with the smallest components; these are typically resistors. Begin at R1 and move through to Rx.
- As each component is installed, check it off the parts list before moving on.
- Confirm the value of each component before it is installed on the PWB. While rare, incorrect packaging or labeling of components can occur. Therefore, verify the marked resistor and capacitor values before installation. If you are a bit rusty on the resistor color codes, check the resistors with an ohmmeter. Capacitors can be checked with a capacitance measurement bridge. While this is perhaps overkill, it is instructive to see the variation in component values, which measurement before installation will reveal. For critical-value components, such as those affecting frequency response (for example, the RIAA curve on the phono preamp), checking components prior to installation may avoid troubleshooting work later.
- Observe proper polarity of devices such as electrolytic capacitors and diodes. A distinctive pad usually denotes the positive terminal in the case of an electrolytic. The silkscreen legend may also provide guidance.
- For devices that may dissipate some amount of heat, such as 5 W resistors, allow extra lead space between the device and the board so as to keep the device above the board. This will minimize heat-caused damage to the PWB and provide for better cooling of the device. For resistors in the 2 to 5 W range, one-half-inch should be sufficient. For higher-power devices, more clearance should be provided.
- Use only the amount of heat and solder necessary to do the job. Once a PWB is damaged by excessive heat, it is very difficult to repair. Excessive solder flux on the board will tend to attract dust, which can lead to arc-over problems at high voltages.
- Inspect each solder connection to be certain that the mounting pad or via has been filled.
- Mount large components last. These typically include barrier strips and high-voltage electrolytic capacitors.

Figure 12.3 shows the initial build of the PWB on the test bench undergoing initial functionality testing. Although it is certainly possible to produce a perfect PWB for a given circuit on the first turn, more often than not, a revision is needed to optimize the design. Looking closely at Figure 12.3b it can be seen that component spacing is tight in some areas of the PWB. The "orange drop" capacitors used in this implementation are rated for 600 V operation. However, given that the B+ voltage at

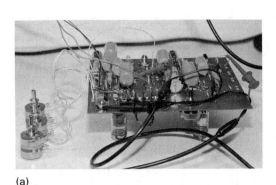

(a)   (b)

FIGURE 12.3 Initial build PWB under test: (*a*) basic measurements taken to confirm proper operation of the circuits, (*b*) close-up view

the input to the board is specified to be approximately 275 V, lower-rated capacitors may be used. In the second build of this PWB, 400 V orange drop capacitors were used, which eliminated the tight spacing.

Once constructed, check the resistance and voltage measurements against the typical values listed in Tables 9.6 and 9.7,[1] respectively.

As detailed in Chapter 9, the value of R1 is specified as 47 kΩ, but the optimum value depends on the type of magnetic pickup used on the turntable. For this reason, mount R1 about a quarter-inch above the board to facilitate replacement if needed.

Circuit details are given in Chapter 9. Remember that the parts list in Table 9.10 is for one PWB. For stereo, the quantities should be doubled, with the exception of R39 (hum balance control), which should be installed on only one board. In this implementation, ganged potentiometers are used for the front panel controls, as suggested in Chapter 9.

A parts list for the chassis is given in Table 12.2. Note that the chassis specified is painted. Establish a ground reference point for the chassis and secure with appropriate hardware to a sanded point on the chassis.

The front panel is shown in Figure 12.4. The panel was done using software provided by Front Panel Express[2]. With this service, the user designs the desired panel and sends the file to the company, which returns the finished product. A variety of finishes and colors are available. For this project, a black panel with white lettering seemed appropriate. Optional finishing touches for the front panel include handles[3] and/or holes for rack mounting. These give the panel a more industrial look that seems to fit with the exposed tube design. If rack mounting is desired, be certain to provide for support at the rear of the chassis.

---

[1] Caution: high voltages are present; be careful.
[2] www.frontpanelexpress.com. There may be other vendors in this space as well.
[3] Included in the parts list in Table 12.2.

TABLE 12.2   Chassis Parts List for the Stereo Preamplifier

Description	Quantity	Manufacturer	Part No.	Allied Stock No.	Notes
Main chassis, 10-in × 17-in × 3-in, steel	1	Hammond	1441-32BK3	806-0534	
Main chassis baseplate, 10-in × 17 in, steel	1	Hammond	1431-30BK3	806-0544	
Front panel, 19-in × 3.5-in	1	Front Panel Express[1]	Custom		
Front panel handles (optional)	2	RAF	8128-832-A-24	219-8007	
Chassis feet	4	Bud	F-7264-A	736-7264	
Large knob	4	Davies	1110	543-1110	Volume/power controls
Handle, side	2	Bud	H-9174-B	736-4374	
RCA connector	10	Neutrik	NF2D-B-0	514-0004	Audio inputs
Power connector, pin	1	Tyco	211767-1	512-8561	Power connector, chassis mount
Power connector, receptacle	1	Tyco	211766-1	512-8869	
Power connector, shell	1	Tyco	206070-8	374-1260	
Power connector, contact pin	10	Tyco	1-66099-5	374-1075	
Power connector, contact socket	10	Tyco	1-66101-9	374-1076	
Terminal strip	4	Cinch	54C	750-6630	
Pilot light	1	Dialight	249-7841-1431-574	511-0276	

[1] See www.frontpanelexpress.com

The back panel input/output connector labels were printed on a self-adhesive polycarbonate overlay by a fabrication house[4] and attached directly to the panel. The design was developed in Adobe Illustrator, and the file was then sent to the vendor, which printed and shipped the overlay. The panel layout is shown in Figure 12.5. A variety of options are available with this type of overlay. Because the intended use was the rear panel, which is rarely seen by most people, no special touches were added.

---

[4] Metalphoto of Cincinnati, www.mpofcinci.com.

FIGURE 12.4   Front panel of the stereo preamplifier

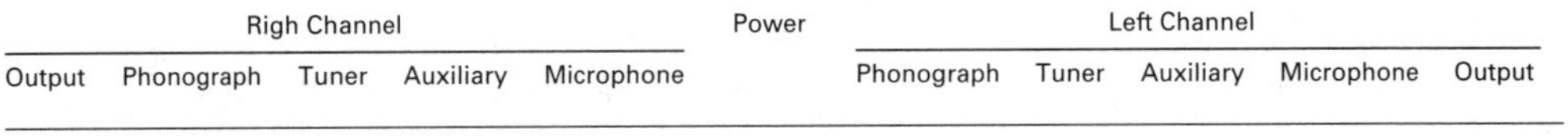

		Righ Channel			Power		Left Channel			
Output	Phonograph	Tuner	Auxiliary	Microphone		Phonograph	Tuner	Auxiliary	Microphone	Output

FIGURE 12.5   Back panel polycarbonate overlay

The finished preamplifier is shown in Figure 12.6. Note the clean design made possible by the two-PWB-board installation. As an additional touch, a decorative Plexiglas shield was mounted above the chassis, as shown in Figure 12.6c. The piece, made to fit above the chassis and around the tubes, was milled to specifications by a local vendor[5] using quarter-inch smoked Plexiglas. This is, of course, an optional element, but it gives an interesting look to the finished unit.

As discussed in Chapter 9, the phonograph and microphone input stage tubes (V1 and V2, respectively) would benefit from the use of shielded sockets to reduce noise pickup. For better or worse, the author decided in this build to use conventional sockets, rather than shielded ones, because of aesthetics. This is certainly not the first time the performance of a consumer audio product was compromised in the name of appearance. As mentioned at the beginning of this book, engineering is an exercise in compromise.

The foregoing notwithstanding, in order to achieve the best (lowest) noise performance from the preamplifier, it is necessary to carefully set the operating points of the preamp stages. A number of variables are involved. The following steps are recommended prior to taking performance measurements:[6]

1. Set the front-panel volume control for approximately quarter-clockwise rotation. Set the front panel bass and treble controls for flat (center) response.
2. Connect an audio voltmeter to the output of the right channel.
3. Switch the input to tuner, and apply a 1 kHz, 0.30 V rms signal to the right channel tuner input. This input level was determined during measurements of the tone-control preamp to yield an output voltage from the stage of 1 V rms into a 1 MΩ load. (See Table 9.8.)

---

[5] TAP Plastics.
[6] Be extremely careful while making adjustments below the chassis, as dangerous voltages exist. Use an insulated adjustment tool; do not place hands near the circuit board or components.

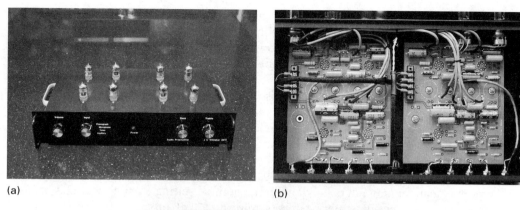

(a)   (b)

(c)

FIGURE 12.6   The stereo audio preamplifier: (*a*) top view, (*b*) bottom view, (*c*) with decorative shield in place

4. Adjust R18 (tuner level control) to the full clockwise position.
5. Readjust the front panel volume control to produce 1 V rms output into a 1 MΩ load at the output terminal of the preamplifier.
6. Repeat Step 3 for the auxiliary input.
7. Adjust R19 (auxiliary level control) to produce 1 V rms into 1 MΩ at the output terminal of the preamplifier. The proper position for R19 should be fully clockwise. Do not readjust the front-panel volume control.
8. Switch the input to phonograph, and apply a 1 kHz, 0.0072 V rms signal to the right channel phonograph input. This signal input level was determined during measurements on the phono preamp stage to yield an output voltage from the stage of 1 V rms (see Table 9.8).
9. Adjust R16 (phono level control) to produce 1 V rms output into a 1 MΩ load at the output terminal of the preamplifier. Do not readjust the volume control.
10. Switch the input to microphone, and apply a 1 kHz, 0.0145 V rms signal to the right channel microphone input. This input signal level was determined during measurements of the mic preamp stage to yield an output voltage from the stage of 1 V rms (see Table 9.8).
11. Adjust R17 (mic level control) to produce 1 V rms output into a 1 MΩ load at the output terminal of the preamplifier.

12. Switch the preamplifier to phonograph, and apply a 1 kHz, 0.0072 V rms signal to the right channel phonograph input. Set the AC voltmeter to read decibels, and adjust for a convenient reference point.
13. Remove the input signal, and connect a shorting connector to the phonograph input.
14. While observing the residual noise level on the voltmeter, adjust R39 (hum balance control) for a minimum reading.
15. Repeat Steps 2 through 14 for the left channel of the preamplifier, with exception of Step 5. Do not readjust the front-panel volume control.

Note that during adjustment of R39, the hum balance control, the optimal setting for one channel may not coincide with the optimal setting for the other channel. The impact of this adjustment can be significant (10 dB or so). In the event that the minimum noise settings do not coincide, split the difference between the two channels.

During adjustment of the tuner/auxiliary input levels, the proper positions for controls R18 and R19 should be nearly identical between channels (fully clockwise). Minor gain variation between channels is normal. For proper balance between channels, it may be necessary to repeat the procedure outlined earlier if the left channel output is noticeably less than the right channel, in which case the left channel gain would be used as the reference point for setting the front-panel volume control.

The foregoing settings should provide a good starting point for use of the preamplifier. Touch-up adjustments using program material can be made later, using the guidelines given in Chapter 9.

After these adjustments have been made, remove power and install the bottom chassis plate. After installing the bottom of the chassis, the measured noise floor should decrease by several decibels.

Measured Performance

Upon completion of the stereo preamplifier, measurements were taken on each channel at an output level of 1 V rms into a load of 1 MΩ. The following test equipment was used:

- Heathkit IG-18 audio generator
- Heathkit IG-1275 lin/log sweep generator
- Heathkit IM-5258 harmonic distortion analyzer
- Heathkit SM-5248 intermodulation distortion analyzer
- Heathkit IM-5238 AC voltmeter
- Heathkit IM-4100 frequency counter
- Heathkit IO-4510 oscilloscope

The measured performance of the stereo preamplifier is documented in Table 12.3. The B+ input voltage was nominally +275 V DC. Measurements were taken at each input position (tuner, auxiliary, phonograph, microphone) for each channel. The tone-control potentiometers were set for their center (flat) response. As detailed in Chapter 9, the flat setting points for the bass and treble controls are best found by using a sweep generator while monitoring the output of the preamplifier with an AC voltmeter. Small variations from the optimal flat response are not unusual between channels. Any deviation of more than about 1 dB may be cause for further investigation.

Comparing the results documented in Table 12.3 with the performance of the individual circuits recorded in Table 9.8, it is evident that the concatenation of the separate circuits results in somewhat poorer performance of the composite system. This is to be expected, and is largely acceptable.

It is important to remember that the performance of any circuit is determined in large measure by the performance of the active devices. In the case of tubes for low-level audio circuits, some suppliers offer specialized services for certain types as a way of optimizing the application. For example, the 7025 (12AX7) can be ordered

TABLE 12.3   Measured Performance of the Stereo Preamplifier

Test	Parameter	Measured Value, Right Channel	Measured Value, Left Channel	Notes
Tuner and Auxiliary Inputs	Input level	0.3 V rms	0.3 V rms	Note 1
	Frequency response	±2 dB, 12 Hz to 20 kHz	±2 dB, 12 Hz to 22 kHz	
	THD	0.6%	0.2%	Note 2
	IMD	0.3%	0.1%	4:1 mix ratio
	Noise, unweighted, (shorted input terminals)	−65 dB	−60 dB	Note 3
Tone Control	Treble control maximum boost	+15 dB	+14 dB	At 20 kHz
	Treble control, maximum cut	−26 dB	−24 dB	At 20 kHz
	Bass control maximum boost	+18 dB	+18 dB	At 20 Hz
	Bass control maximum cut	−19 dB	−18 dB	At 20 Hz
Phonograph Input	Input level	0.0072 V rms	0.0072 V rms	
	Frequency response	±2 dB, 20 Hz to 20 kHz	±2 dB, 20 Hz to 20 kHz	Note 4
	THD	0.6%	0.6%	Note 2, note 5
	IMD	–	–	Note 6
	Noise, unweighted, (shorted input terminals)	−67 dB	−64 dB	Note 3

(continued)

TABLE 12.3   Measured Performance of the Stereo Preamplifier (*Continued*)

Test	Parameter	Measured Value, Right Channel	Measured Value, Left Channel	Notes
Microphone Input	Input level	0.0145 V rms	0.0145 V rms	
	Frequency response	±3 dB, 20 Hz to 20 kHz	±3 dB, 20 Hz to 20 kHz	
	THD	0.4%	0.3%	Note 2
	IMD	0.8%	0.6%	4:1 mix ratio
	Noise, unweighted, (shorted input terminals)	−65 dB	−65 dB	Note 3

Notes:

1 For 1 V rms output at 1 kHz into 1 MΩ.

2 THD measurements were taken at 20 Hz, 100 Hz, 1 kHz, 10 kHz, and 20 kHz unless otherwise noted. The worst (highest) measurement is the one quoted.

3 Noise measurements were taken with all covers in place. Note that the left channel noise is a couple of decibels higher than the right channel in most measurements. This is the result of different optimal setting points for the hum balance control. In an ideal world, the optimal point for both amplifiers would coincide. As a practical matter, the best that can be hoped for is a compromise point that produces acceptable results. A different set of tubes may have a different optimal point.

4 Relative to the Recording Industry Association of America (RIAA) equalization curve. See Table 9.9.

5 By necessity, THD measurements on the phono preamp took into considerations the RIAA frequency-response curve. If the same input level was used for all measurements, the preamp would overload on the low end and be buried in the noise at the high end. The following procedure was used: 1) a reference output level from the preamp was established at 1 kHz; 2) for measurements at 20 Hz and 100 Hz, the input level was adjusted so that the output from the preamp matched the level at 1 kHz; 3) for the measurement at 10 kHz, the input level was maintained at the 1 kHz level and the gain of the distortion analyzer was increased to compensate for the response rolloff; and 4) for the measurement at 20 kHz, the gain of the analyzer was fixed at the position for the 10 kHz measurement and the input level was increased to satisfy the reference input level of the instrument (about 9 dB increase). This is the same procedure described in Table 9.8.

6 No IMD measurements were taken on the phono preamp, as the frequency-response curve of the RIAA transfer characteristic would invalidate the measurement.

from at least one online vendor[7] as: 1) standard functional test, 2) high gain, 3) low noise and microphonics, or 4) matched set. For the preamplifier, it would probably be a good idea to consider the low noise selection service, since any reduction in noise— particularly for the phono and mic stages—will be beneficial to overall performance.

The measurements recorded in Table 12.3 were taken with tubes ordered without any special selection. The intent was to establish a baseline of nominal performance points that, with common components, could be met. Improved or specially selected components should result in improved overall performance.

---

[7] www.tubedepot.com; there may be other vendors offering this service as well. The cost difference for the various selection options is typically small.

After completing any project, it is useful to consider items that the builder would do differently on the next project. In the case of the stereo preamplifier, the only item on which the author had second thoughts was the use (or in this case nonuse) of shielded tube sockets for the phonograph and microphone preamp devices (and perhaps the other tubes as well). Without the shields, a noise floor of –60 dB or better (lower) was achieved using stock tubes. This is not great, but not bad either. For listening at home where computers are running and kids are running around, finding a point in time that a –60 dB noise floor would bother the listener is likely difficult; having said that, different builders understandably have different goals and objectives. Food for thought.

25 W Stereo Amplifier

This project consists of two identical 12.5 W amplifiers as described in Chapter 10 (Figure 10.4). The circuits are built on the PWB design outlined in Chapter 10, along with some finishing touches on the chassis.

For nearly any high-performance audio amplifier, the limiting factor is usually the transformers—not just because of the cost (which can be 50 percent of the bill of materials), but also because of the physical size and weight of the devices. In deciding on a power amplifier design, it is necessary, therefore, to look at the system from the transformers back to the other active and passive components. Complicating this process is the generally limited selection of power transformers and output transformers available commercially. Such was not always the case, of course, when vacuum tubes were the primary devices that made consumer electronics work. The large selection of transformers available back in the day, regrettably, is somewhat narrow now. Having said that, audio enthusiasts have kept the high-end transformer business in business; for that we can all be thankful.

The circuit description given previously will not be repeated here. Rather, details will be limited to construction techniques and final performance measurements. Remember that the parts list in Table 10.7 is for one channel. For a stereo system, the quantities shown should be doubled. The exception here is the power supply components, which should be installed on only one board. As stated in Chapter 10, the power supply has been sized to drive two channels. Installation details for the 25 W stereo amplifier are provided in Table 12.4. For the purposes of description, the right channel is designated as the *primary channel* and the left channel is designated as the *secondary channel.* Functionally, the two are identical, of course.

Note that transformer T3 shown in Figure 10.4 is intended to drive the external preamplifier heaters. The 6.3 V secondary windings are connected through a 6 A fuse to the back panel connector, as detailed in Table 12.5.

For this implementation, the amplifier was built on a large chassis with exposed tubes to provide a showcase appearance. Because of the weight of the transformers, a heavy-gauge steel chassis was used. A parts list for the chassis components is given in Table 12.6.

Prior to mounting the PWBs onto the main chassis, each assembly was checked for proper resistance measurements, as detailed in Table 10.3. This step is very

TABLE 12.4 Wiring Considerations for the Primary and Secondary Channels

Consideration	Primary Channel	Secondary Channel
PWB components	Install all components specified in Figure 10.4 for the PWB	Install only the amplifier components on the PWB. Do not install R14, R16, R18, R21, R22, C9, C10, C11, C12, C13, D1, D2, D3, CR1, CR2, CR3, VR2
Power transformer T2	Install connections as detailed in Table 10.8[1]	No connections from T2, except for the 6.3 V heater winding to Term 5 and Term 6. These connections can be made from the appropriate points on R13 (hum balance control).
Chassis components	Install all components specified in Figure 10.4	No chassis components other than the output transformer and volume control.
Power connections	Terminal 1 on PWB to back panel power connector through a 0.5 A fuse (auxiliary B+ supply)[2]	Terminal 1 not connected
	Terminal 2 to chassis ground	Terminal 2 to chassis ground
	Terminal 3 of primary PWB to terminal 3 of secondary PWB (bias supply)[3]	
	Terminal 4 of primary PWB to terminal 4 of secondary PWB (B+ supply)	

[1] Caution: Wiring color codes may vary from one transformer to the next. Consult the manufacturer.
[2] The back panel connector pinout is given in Table 12.5.
[3] The bias adjustment on the primary channel PWB sets the operating point for both channels.

TABLE 12.5 Back Panel Power Connector Pinout

Pin No.	Function	Connects To	Notes
1	Ground	Terminal 2 of primary PWB	
2	Pilot lamp	PL1 through 1 kΩ, 0.5 W, resistor[1]	117 V AC
3	Pilot lamp	PL1	117 V AC
4	Auxiliary B+	Terminal 1 of primary PWB through a 0.5 A inline fuse	Approximately +275 V DC
5	Filament	T3 secondary winding through a 6 A inline fuse	6.3 V AC for heaters
6	Filament	T3 secondary	6.3 V AC for heaters
7			Not used
8			Not used
9	Ground	Chassis ground	

[1] The 1 kΩ resistor is placed on the hot side of the 117 V AC connection to PL1. Its function is to limit damage in the event of a wiring error or other connection problem wherein the pilot lamp wires become short-circuited. In such a case, the resistor would burn out and break the circuit.

TABLE 12.6   Chassis Parts List for the 25 W Stereo Amplifier

Description	Quantity	Manufacturer	Part No.	Allied Stock No.	Notes
Main chassis, 14-in × 17-in × 3-in, steel	1	Hammond	1441-38BK3	806-1396	
Main chassis baseplate, 14-in × 17 in, steel	1	Hammond	1431-38BK3	806-1426	
Front panel, 19-in × 3.5-in	1	Front Panel Express[1]	Custom		
Front panel handles (optional)	2	RAF	8128-832-A-24	219-8007	
Chassis feet	4	Bud	F-7264-A	736-7264	
Large knob	2	Davies	1110	543-1110	Volume/power controls
Handle, side	2	Bud	H-9174-B	736-4374	
RCA connector	2	Neutrik	NF2D-B-0	514-0004	Audio inputs
4-term barrier strip	2	Waldom Molex	38720-3204	607-0065	Speaker outputs
Power switch	1	NKK Switches	WB12T-DA	870-0669	Front panel power switch
Power cord	1	Alpha Wire	615 BK078	663-7020	
Power cord strain relief	1	Thomas & Betts	2672	534-0846	
Power connector, socket	1	Tyco	211767-1	512-8872	Power connector, chassis mount
Power connector, pin	1	Tyco	211766-1	512-8873	
Power connector, shell	1	Tyco	206070-8	374-1260	
Power connector, contact pin	10	Tyco	1-66099-5	374-1075	
Power connector, contact socket	10	Tyco	1-66101-9	374-1076	
Terminal strip	4	Cinch	54C	750-6630	
Fuse, 6 A, slo-blow	2	Bussmann	MDL-6-R	740-0752	
Fuse, 0.5 A, slo-blow	2	Littelfuse	0313.500HXP	845-0307	
Fuse block	2	Bussman	S-8002-1-R	740-0772	For external filament and plate circuits
1 kΩ, 0.5 W, resistor	1	Ohmite	OL1025E	296-4772	For external pilot lamp circuit

[1] See www.frontpanelexpress.com.

helpful in identifying board wiring errors, including installation of the wrong value of component. Be aware that certain resistance benchmarks, notably the plate and screen pins of the output tubes, will show an open circuit if the output transformer is not connected. Note also that the control grid reading for the input pentode will be an open circuit to ground if the volume control is not installed, and the heater connections will show an open circuit if R13 (hum balance control) is not installed.

The scope of resistance readings that can be done on the secondary channel PWB will be quite limited, since there are no power supply components installed on the board. One way around this limitation is to cross-connect the two boards (primary PWB terminal 1 to secondary PWB terminal 1, primary PWB terminal 2 to secondary PWB terminal 2, primary PWB terminal 3 to secondary PWB terminal 3, and primary PWB terminal 4 to secondary PWB terminal 4). Be certain to remove these temporary jumpers prior to installation of the boards.

After both PWBs and the related chassis components have been installed, repeat the checkout steps detailed in Chapter 10, in particular the resistance measurements in Table 10.3 and the voltage measurements in Table 10.4.[8] Remember the caution about possible motorboating/oscillation during initial startup.

Note that the power transformer specified in Table 10.7 for this amplifier (Hammond 274BX) accommodates line input voltages of 115 V AC and 120 V AC. For operation at 240 V AC, a different power transformer must be used, such as the Hammond 374BX.

Voltage readings were taken on the two amplifier channels with all tubes installed, and they closely followed the nominal readings listed in Figure 10.4. It is worth noting, however, that considerable differences (as much as 15 to 20 percent) for the triode section of the 7199 phase-splitter stage were observed from one channel to the other. These were the result of differences in the 7199 devices themselves (the voltage variations moved with the tubes). While both tubes operated satisfactorily, it is advisable to use devices from the same manufacturer for the 7199 input stage of each channel, as this may reduce device variability. The voltages listed in Figure 10.4 for the 7199 stages were for an amplifier using the JAN/Philips new old stock (NOS) device.

During testing, it was found that different 7199 tubes exhibit different (sometimes significant) noise characteristics. In one case during measurements on this amplifier, a 15 dB noise floor variation was observed from one device to another. Both tubes otherwise provided identical performance.

A check of the operating voltages listed in Figure 10.4 against the device specifications given in Chapter 7 show both tube types operating well within their maximum ratings. Of note is the 6973 beam power tube, which for this mode of operation (screen tap and fixed bias), is right in line with the recommended values given in Table 7.9.

During the time between the application of power and completion of filament warm-up, the voltage at the output of the power supply stage will rise to +450 V DC or perhaps slightly higher. As the filaments warm up and the amplifiers begin to draw

---

[8] Caution: High voltages are present during this step.

FIGURE 12.7   Front panel for the 25 W stereo amplifier

	Right Channel	Power	Left Channel	
Input	Speaker Output (GND, 4Ω, 8Ω, 16Ω)	High Voltage Do Not Probe	Speaker Output (16Ω, 8Ω, 4Ω, GND)	Input

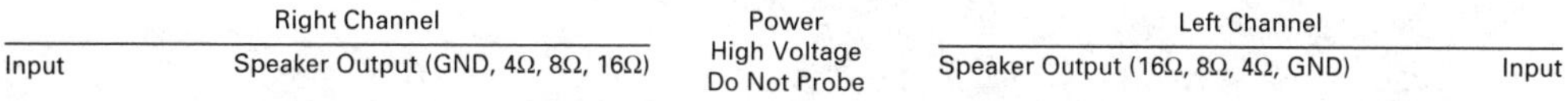

FIGURE 12.8   Back panel polycarbonate overlay for the 25 W stereo amplifier

current, the B+ voltage decreases to approximately +350 V DC, which is the nominal operating point of the supply. Filter capacitors in the power supply and amplifier circuits are rated for operation at 450 V DC. The so-called *surge voltage* rating of these devices is somewhat higher than 450 V, providing a margin of operating safety. Nevertheless, it is helpful to limit the turn-on voltage excursion to some lower level. To accomplish this, two zener diodes, each rated for 200 V at 5 W[9], were connected in series to yield an operating point of 400 V. The diode pair was connected between ground and Terminal 1 on the right channel amplifier PWB. This has the effect of holding the maximum B+ voltage to something less than +450 V DC. When the zener diode pair conducts, it draws current through R14 and loads the power supply until the tubes have warmed up. Once the B+ voltage falls below about +400 V DC, the zener pair stops conducting, and from that point on is essentially out of the circuit.

The front panel for the 25 W stereo amplifier was produced in the same manner as described previously for the stereo preamplifier. The finished panel is shown in Figure 12.7. The back panel legend was likewise done in the same manner as for the preamplifier (Figure 12.8).

The completed unit is documented in Figure 12.9. Because nearly all of the components are mounted on the PWBs, construction of the amplifier is relatively straightforward. As with the preamplifier, a decorative Plexiglas shield was added, in this case over the power and output transformers.

After operation of this amplifier for several weeks, the author tried a different approach with the Plexiglas shield. In keeping with the design for the stereo preamplifier, a quarter-inch smoked Plexiglas shield was milled to go around the tubes and the transformers. While it was a complex (and rather expensive) cut, the end result was very attractive. This approach also had the benefit of reducing heating of the power transformer, which after long periods of operation can become quite warm.

---

[9] ON Semiconductor, Part No: 1N5388BG, Newark Electronics stock #70K6988.

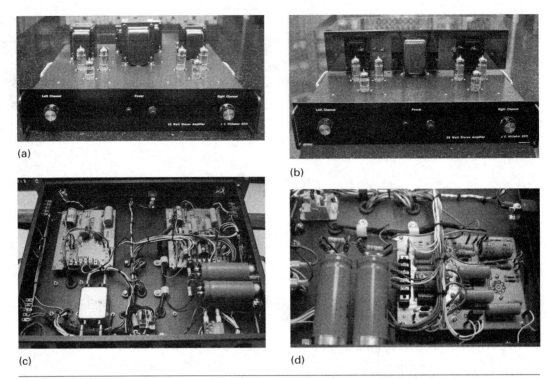

FIGURE 12.9 The 25 W stereo audio amplifier: (*a*) top view, (*b*) with decorative cover in place, (*c*) bottom view, (*d*) detail view of the right channel PWB

The chassis used was painted, and the transformers were shipped painted. To ensure the case of each transformer was grounded, the paint was sanded off one mounting pad of each transformer and a ground lug attached. This is important for safety reasons. While quite rare, transformer failures can occur. One failure mode can be a shorting of one of the windings to the transformer case. By grounding the case, such a failure would open the circuit breaker and prevent dangerous voltages above the chassis in a fault mode.

Measured Performance

Upon completion of the amplifier, measurements were taken at output levels of 1 W, 5 W, 10 W, and 12.5 W, the maximum rated output of each channel. Measurements were made on one channel at a time; the input terminals of the channel not being measured were shorted. The following test equipment was used:

- Heathkit IG-18 audio generator
- Heathkit IG-1275 lin/log sweep generator
- Heathkit IM-5258 harmonic distortion analyzer

- Heathkit SM-5248 intermodulation distortion analyzer
- Heathkit IM-5238 AC voltmeter
- Heathkit IM-4100 frequency counter
- Heathkit IO-4510 oscilloscope
- Heathkit SD-4850 digital storage oscilloscope

For the measurements, an 8 Ω 20 W noninductive resistor was connected to the output terminals of each channel. The measured performance of the 25 W stereo amplifier is given in Table 12.7.

There were no real surprises in the performance measurements, as the amplifier was well characterized in the tests documented in Chapter 10. More than anything else, this was a test of the output transformer, which was different from the one used in the Chapter 10 test bed trial. Both transformers were Hammond units with similar specifications, the major difference being the power rating (40 W as opposed to 20 W). There were no appreciable differences in overall performance, with the exception of

TABLE 12.7   Measured Performance of the 25 W Stereo Power Amplifier

Conditions	Parameter	Measured Value, Right Channel	Measured Value, Left Channel	Notes
Measurements taken at 1 W power output	Input voltage	0.31 V rms	0.29 V rms	Volume control fully clockwise
	Frequency response	±1 dB, 20 Hz to 65 kHz	±1 dB, 20 Hz to 55 kHz	Note 1
	THD	0.3%	0.3%	Note 2
	IMD	0.38%	0.42%	4:1 mix ratio
	Noise unweighted, (shorted input terminals)	−81 dB	−85 dB	Volume control fully clockwise
Measurements taken at 5 W power output	Input voltage	0.64 V rms	0.68 V rms	Volume control fully clockwise
	Frequency response	±1 dB, 15 Hz to 50 kHz	±1 dB, 10 Hz to 45 kHz	
	THD	0.8%	0.8%	Note 2
	IMD	1.2%	1.28%	4:1 mix ratio
	Noise, unweighted, (shorted input terminals)	−87 dB	−92 dB	Volume control fully clockwise

(continued)

TABLE 12.7 Measured Performance of the 25 W Stereo Power Amplifier (*Continued*)

Conditions	Parameter	Measured Value, Right Channel	Measured Value, Left Channel	Notes
Measurements taken at 10 W power output	Input voltage	0.93 V rms	0.93 V rms	Volume control fully clockwise
	Frequency response	±1 dB, 17 Hz to 45 kHz	±1 dB, 16 Hz to 40 kHz	
	THD	1.2%	1.2%	Note 2
	IMD	1.5%	1.3%	4:1 mix ratio
	Noise, unweighted, (shorted input terminals)	–94 dB	–95 dB	Volume control fully clockwise
Measurements taken at 12.5 W power output	Input voltage	1.1 V rms	1.1 V rms	Volume control fully clockwise
	Frequency response	±1 dB, 20 Hz to 40 kHz	±1 dB, 20 Hz to 35 kHz	
	THD	1.3%	1.7%	Note 2, note 3
	IMD	1.6%	1.9%	4:1 mix ratio
	Noise, unweighted, (shorted input terminals)	–92 dB	–95 dB	Volume control fully clockwise, note 4

Notes:

1 The low-frequency response of the amplifier at 1 W power output exhibits a bump of about +3 dB at approximately 12 Hz in both channels. The response of the amplifier goes well below the 20 Hz number quoted, but it is not within the ±1 dB range.

2 THD measured at 30 Hz, 100 Hz, 1 kHz, 10 kHz, and 20 kHz unless otherwise noted. The value recorded is the highest reading. The lower point for THD mesurement was set at 30 Hz because that is the lower operating range quoted by the manufacturer for the output transformers.

3 The distortion measurement of the left channel at 12.5 W was below 1 percent at all test frequencies, except for 20 kHz. It was possible to lower this value through adjustment of the bias, but at the expense of raising the distortion slightly in the right channel. Given the small deviation in THD between channels, the distortion performance of the right channel was optimized through adjustment of the bias and no additional changes were made to the bias setting during measurements on the left channel. Remember the cautions described in Chapter 10 regarding adjustment of bias. The author's recommendation is to find the optimal bias setting and then back it off (make the bias more negative) by about 5 V. This will avoid the problem of shortened tube life due to high idling current. As a general observation, if the power transformer runs hot, the bias is set too low.

4 Note the exceptional noise performance of the left channel amplifier. Recall that the two PWBs are identical; however, the right channel PWB includes the power supply components and the left channel board does not. While the differences in noise performance between the two channels are small (and may be the result of various factors), the measurements suggest an implementation wherein the power supply components are placed on a separate PWB. The drawbacks to this approach include additional layout complexity and increased cost. In any event, the ground plane design of the PWB appears to serve the intended purpose. Noise measurements below –90 dB are impressive, and both channels meet that number.

the low-frequency bump in response at low power levels (see Table 12.7, Note 4). The variance here is minor and can largely be ignored. The clipping point of the amplifier was observed at 15.5 W into 8 Ω. Clipping of the positive and negative traces was symmetrical.

The square wave response of the amplifier was measured at 2 W output (roughly equal to the "average listening level") and 15 W (the maximum output capability of the circuit[10]) into an 8Ω load. The measurements are shown in Figure 12.10. The traces were taken from the right channel amplifier; performance of the left channel was essentially identical. The following observations can be made:

- **At 2 W power output**   1) A fair amount of low-frequency tilt was observed at low frequencies (100 Hz and below). This may be the result of phase shift through the amplifier. 2) The response at 1 kHz looks quite good. 3) The response at 10 kHz is respectable, with no signs of instability.
- **At 15 W power output**   1) Considerable low-frequency tilt was observed at low frequencies (100 Hz and below). Clearly the amplifier is running at the edge of its capabilities. 2) The trace at 1 kHz looks good, with slight ringing on the leading edge. 3) At 10 kHz, the trace shows significant ripple, which is not entirely unexpected at the operating limit. Probing the response of any circuit at the limits of its capabilities is instructive, if only to reaffirm that operating below that level is a good practice.

Restating the obvious, operation of this amplifier (or any other amplifier) at its maximum operating point has performance penalties. However, for listening at modest levels, the amplifier performs quite well.

Readers will recall the option, detailed in Chapter 10, to substitute a 6U8A (with suitable socket connection changes) for the 7199 input tube. Given the excellent performance of the amplifier with the 7199 and the continued availability of the device (at present, anyway), this option was not tried. The 6U8A will no doubt seem like a better idea when the 7199 is really hard to find—and consequently really expensive. For now, the author's recommendation is to stay with the 7199.

Since the shelf life of a receiving tube is essentially unlimited, it is good practice to purchase at least one spare tube for each type used in a project. In the case of the power output tubes, which for best performance should be matched, a pair of tubes is recommended.

After completing this project, the author has no suggested improvements for the next version of this amplifier. The performance of the 25 W stereo amplifier met expectations in all categories—and with regard to noise performance, exceeded expectations. It is very satisfying to finish a project and not want to change it!

---

[10] Note that this output level is above the rated power output of 12.5 W. Discussion of the 15 W versus 12.5 W output capability of this amplifier can be found in Chapter 10.

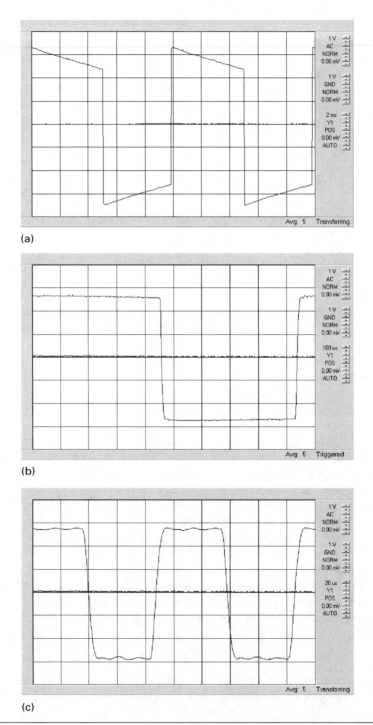

(a)

(b)

(c)

FIGURE 12.10 Square wave response of the amplifier: (*a*) 100 Hz at 2 W, (*b*) 1 kHz at 2 W, (*c*) 10 kHz at 2 W (*Continued*)

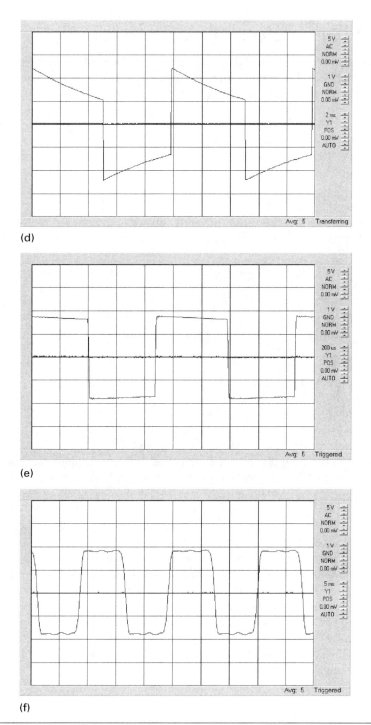

(d)

(e)

(f)

FIGURE 12.10 Square wave response of the amplifier: (*d*) 100 Hz at 15 W, (*e*) 1 kHz at 15 W, (*f*) 10 kHz at 15 W

50 W Stereo Power Amplifier

This unit consists of two identical 25 W power amplifiers as described in Chapter 11 (Figure 11.3). The power supply is based on the circuits described in Chapter 8. The exceptional performance of this amplifier can be attributed to the following:

- Use of inverse feedback from the voice-coil winding of the output transformer to the cathode of the input amplifier.
- Use of high-performance output transformers, conservatively rated for the application.
- Use of a robust plate supply for the beam power tubes in the output stage.
- Operation of all heaters at a positive voltage with respect to ground and the use of a balancing adjustment in the heater supply circuit, both intended to minimize hum.
- Provisions for adjustment of the control grid bias of the output beam power tubes.

The circuit description and accompanying tube data contained in previous chapters will not be repeated here.

The 50 W stereo amplifier was built on a large chassis with exposed tubes to provide a showcase appearance. Because of the weight of the transformers, a heavy-gauge chassis was used. The weight of the amplifier is largely determined by the four transformers as follows:

- Output transformer = 7 lbs each
- Power transformer = 10 lbs
- Power supply choke = 8 lbs

The total transformer weight is about 32 lbs.

The power supply used was identical to the final regulated circuit design shown in Figure 8.6, with the exception of the "balance" rheostat for the 6080 series regulator tube. The specified device in Table 8.7 for R5 is a 50 Ω, 25 W rheostat. This device is intended to adjust for balance between the two triodes of the 6080. As noted in Chapter 8, a 12 W device would suffice, but was not specified in favor of a larger (25 W) rheostat in order to maintain sufficient operating voltage margin. (The 12 W device is rated for operation at up to 305 V, whereas the 25 W device is rated for 750 V.) After a considerable number of hours of operation, the author observed that adjustment of R5 was rarely necessary. Being mindful of limited chassis space on the 50 W stereo power amplifier, the rheostat was eliminated in favor of an adjustable fixed resistor.[11] The tap position is changed by loosening the set-screw, repositioning the tap, and then tightening the set-screw. With this device, of course, adjustments can only be made when power is off, but this was not considered a major issue since adjustment was so rarely needed.

---

[11] 50 Ω, 12 W, wire-wound adjustable tubular resistor, Ohmite #D12K50RE, Allied stock #296-4432. Various mounting options are available.

The following procedure is recommended for the adjustable tubular wire-wound resistor option:

1. Remove all power from the chassis, and ensure that all power supply filters are completely discharged. As a precaution, connect a clip lead from ground to one of the terminals on the adjustable resistor while making changes. *Be certain to remove the clip lead before reapplying power.*
2. Position the R5 tap at center position and tighten.
3. Apply power and carefully measure the voltage to ground at each end of the resistor with a typical load connected to the output of the regulated supply. The voltages should match within ±1 V.
4. If the voltages are not properly matched, remove power and discharge all power supply capacitors, and then reposition the tap. Once secured, reapply power and check the voltages again.

For most applications, simply setting the tap at the center point should be sufficient for proper operation. Note that settings of R5 off the center position are an indication of differential output from the 6080 series tube triode stages. At some point, replacement of the tube is a better option than changing the tap to compensate.

For the 50 W stereo power amplifier, the power supply uses the capacitor-input filter option shown in Figure 8.6. Devices C1, C2, R1, R2, and R23 should therefore be installed. Table 8.7 lists typical output voltages from the power supply at various load conditions. Items that will impact the B+ supply voltage include, but are not limited to: 1) the applied primary voltage, 2) primary input tap chosen, and 3) bias setting on the output tubes. In general, a B+ voltage of about +450 V DC can be assumed from the supply shown in Figure 8.6 with the capacitor-input option and driving two amplifier channels.

Figure 12.11 is a copy of the Figure 11.3 amplifier, but with the typical operating voltages assuming a B+ input supply of +450 V DC. The components of the amplifier, notably electrolytic capacitors, are rated to handle this voltage (see Table 11.7).

A parts list for the chassis components is given in Table 12.8.

Once all components have been installed, repeat the steps in Chapter 11 in the section "Initial Checkout."[12]. Resistance and voltage measurements should be taken on each amplifier channel using Table 11.3 and Table 11.4, respectively, as benchmarks. Note that because a different method of hum balance on the heaters is used in this design relative to the original RCA circuit in Figure 11.1, the heater pins will typically show 4 kΩ to ground during the "no power, no tubes" test, and a voltage of about +50 V DC during the "power on but no tubes" test. Remember the caution about motorboating or oscillation upon initial startup.

The chassis elements described previously for the stereo preamplifier and 25 W stereo amplifier were used for the 50 W stereo power amplifier, including the etched front panel[13] (Figure 12.12) and rear panel polycarbonate overlay (Figure 12.13).

---

[12] Caution: dangerous voltages are present during portions of this process.
[13] In the initial build of the amplifier, a prototype panel was used, which differs only slightly from the "final" panel design shown in Figure 12.11. (At $100 each, zero waste is a good policy!)

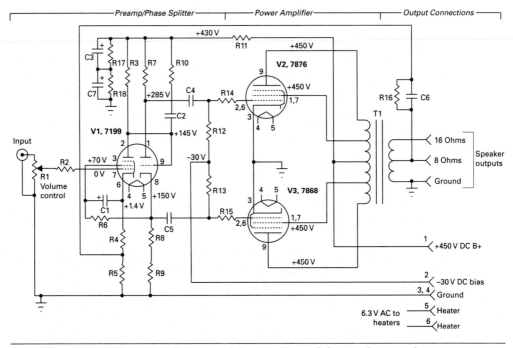

FIGURE 12.11 Schematic diagram of the 25 W amplifier with typical operating voltages shown for an input B+ of 450 V DC

TABLE 12.8 Chassis Parts List for the 50 W Stereo Power Amplifier

Description	Quantity	Manufacturer	Part No.	Allied Stock No.	Notes
Main chassis, 14-in × 17-in × 3-in, steel	1	Hammond	1441-38BK3	806-1396	
Main chassis baseplate, 14-in × 17 in, steel	1	Hammond	1431-38BK3	806-1426	
Front panel, 19-in × 3.5-in	1	Front Panel Express[1]	Custom		
Front panel handles (optional)	2	RAF	8128-832-A	219-8007	
Chassis feet	4	Bud	F-7264-A	736-7264	
Large knob	2	Davies	1110	543-1110	Volume/power controls
Handle, side	2	Bud	H-9174-B	736-4374	
RCA connector	2	Neutrik	NF2D-B-0	514-0004	Audio inputs

(continued)

TABLE 12.8 Chassis Parts List for the 50 W Stereo Power Amplifier (*Continued*)

Description	Quantity	Manufacturer	Part No.	Allied Stock No.	Notes
4-term barrier strip	2	Waldom Molex	38720-3204	607-0065	Speaker outputs
Power switch	1	NKK Switches	WB12T-DA	870-0669	Front panel power switch
Power connector, socket	1	Tyco	211767-1	512-8872	Power connector, chassis mount
Power connector, pin	1	Tyco	211766-1	512-8873	
Power connector, shell	1	Tyco	206070-8	374-1260	
Power connector, contact pin	10	Tyco	1-66099-5	374-1075	
Power connector, contact socket	10	Tyco	1-66101-9	374-1076	
Power cord strain relief	1	Thomas & Betts	2672	534-0846	
Terminal strip	8	Cinch	54C	750-6630	
Fuse, 0.5 A, slo-blow	2	Littelfuse	0313.500HXP	845-0307	
Fuse, 6 A, slo-blow	2	Bussmann	MDL-6-R	740-0752	
Fuse block	2	Bussman	S-8002-1-R	740-0772	For external filament and plate circuits
1 kΩ, 0.5 W, resistor	1	Ohmite	OL1025E	296-4772	For external pilot lamp circuit

[1] See www.frontpanelexpress.com.

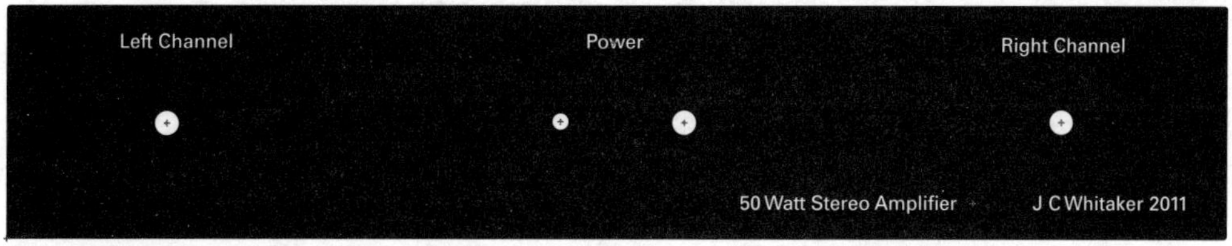

FIGURE 12.12 Front panel of the 50 W stereo power amplifier

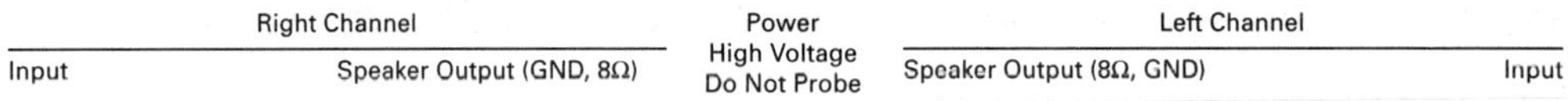

	Right Channel		Power	Left Channel	
Input	Speaker Output (GND, 8Ω)		High Voltage Do Not Probe	Speaker Output (8Ω, GND)	Input

FIGURE 12.13 Rear panel polycarbonate overlay for the 50 W stereo power amplifier

Note that for the rear panel, a single 8Ω speaker output tap is assumed for each channel, as would be the case for the Hammond 1650H output transformer. In the case of the Hammond 1650HA, the panel layout should be modified to include text describing taps for 4, 8, and 16 Ω.

Back panel power connections for the 50 W stereo amplifier are shown in Table 12.9. Note that the pinout for this amplifier provides the same functionality as the 25 W stereo amplifier, allowing the stereo preamplifier (or another unit) to be used with either amplifier.

Due to cost considerations and available time, the author built the 50 W stereo amplifier using hand-wired terminal strips. This reduced the project cost considerably; three PWBs would be needed for this project, at a cost (in small quantities) of more than $500 just for the circuit boards. The time to completion was also a consideration. As documented previously in this chapter for the preamplifier and the 25 W stereo amplifier, there are a number of benefits to the PWB approach. Assembly time is not

TABLE 12.9 Back Panel Power Connector Pinout

Pin No.	Function	Connects To	Notes
1	Ground	Terminal 4 of power supply	
2	Pilot lamp	PL1 through 1 kΩ, 0.5 W, resistor[1]	117 V AC
3	Pilot lamp	PL1	117 V AC
4	Auxiliary B+	Terminal 3 of power supply through a 0.5 A inline fuse	+275 V DC, regulated
5	Filament	Terminal 8 of power supply board through a 6 A inline fuse	6.3 V AC for heaters
6	Filament	Terminal 9 of power supply	6.3 V AC for heaters
7			Not used
8			Not used
9	Ground	Chassis ground	

[1] The 1 kΩ resistor is placed on the hot side of the 117 V AC connection to PL1. Its function is to limit damage in the event of a wiring error or other connection problem wherein the pilot lamp wires become short-circuited. In such a case, the resistor would burn out and break the circuit.

TABLE 12.10   PWB Wiring Considerations

Power Supply	Right Channel Amplifier	Left Channel Amplifier	Back Panel Connector
Terminal 1 (supply output)	Terminal 1	–	–
Terminal 2 (supply output)	–	Terminal 1	–
Terminal 3 (regulated supply output)	–	–.	Pin 4 through 0.5 A inline fuse
Terminal 4 (ground)	Terminal 4	Terminal 4	Pin 1
Terminal 5 (6.3 V heater)	Terminal 5	Terminal 5	
Terminal 6 (6.3 V heater)	Terminal 6	Terminal 6	
Terminal 7 (bias supply)	Terminal 2	Terminal 2	
Terminal 8 (6.3 V filament)	–	–	Pin 5 through a 6 A inline fuse
Terminal 9 (6.3 V filament)	–	–	Pin 6

one of them, as working with the circuit boards requires considerably more time than hand-wiring terminal strips. Having said that, it is fair to point out that the benefits of PWB-based construction are rarely achieved in one-off projects. Rather, longer product runs can realize considerable efficiencies. Also, making the first PWB is always time-consuming, as additional testing is done to confirm the design. Building board #2 can take half the time to build board #1, and board #3 is quicker still.

As detailed in Chapters 8 and 11, PWBs were designed for both the power supply and power amplifier.[14] For such implementations, the power supply PWB can be installed with or without the +275 V regulated output. The two 25 W amplifier channels are identical. Interconnections among the three PWBs are described in Table 12.10. Connections from power supply transformers T1, T2, and T3 are detailed in Table 8.8. Connections from the output transformers to the right and left channel amplifiers are detailed in Table 11.7. Note that wiring color codes vary; consult the manufacturer's literature.

The completed unit is documented in Figure 12.14. For any project, hindsight is always 20/20. In the case of this project, the author would probably not include the regulated power supply option in the next build of this amplifier. While the regulator performs quite well, the chassis is crowded with the additional components (notably the three tubes) used in the circuit. As a practical matter, a voltage divider and filter off the B+ supply would be more than adequate for an auxiliary unit, such as the stereo preamplifier. Also, the heat given off by the amplifier is significant, and so reducing the tube count would help to reduce the "space heater" effect.

---

[14] See the website for implementation details.

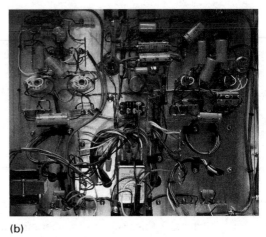

(a) (b)

FIGURE 12.14   The 50 W stereo audio amplifier: (*a*) top view, (*b*) bottom view

Measured Performance

Upon completion of the 50 W stereo amplifier, measurements were taken at output levels of 1 W, 5 W, 10 W, 20 W, and 25 W, the maximum rated output of each channel. Measurements were made on one channel at a time; the input terminals of the channel not being measured were shorted. The following test equipment was used:

- Heathkit IG-18 audio generator
- Heathkit IG-1275 lin/log sweep generator
- Heathkit IM-5258 harmonic distortion analyzer
- Heathkit SM-5248 intermodulation distortion analyzer
- Heathkit IM-5238 AC voltmeter
- Heathkit IM-4100 frequency counter
- Heathkit IO-4510 oscilloscope
- Heathkit SD-4850 digital storage oscilloscope

For the measurements, an 8 Ω 30 W noninductive resistor was connected to the output terminals of each channel. The measured performance of the 50 W stereo power amplifier is given in Table 12.11.

The performance of this amplifier was virtually identical to the measurements recorded in Chapter 11. No real surprise here, since the major components are all the same. Furthermore, the test bed chassis was repurposed for the 50 W stereo power amplifier, and so the component layout was quite similar to the circuit built in Chapter 11. The only significant performance difference was an improvement in the noise floor, attributable to better separation of heater lines from the input circuits.

Square wave performance of the amplifier is documented in Figure 12.15 for input frequencies of 100 Hz, 1 kHz, and 10 kHz at power output levels of 2 W, 15 W,

TABLE 12.11   Measured Performance of the 50 W Stereo Power Amplifier

Conditions	Parameter	Measured Value, Right Channel	Measured Value, Left Channel	Notes
Measurements taken at 1 W power output	Input voltage	0.12 V rms	0.12 V rms	Volume control fully clockwise
	Frequency response	±1 dB, 6 Hz to 60 kHz	±1 dB, 5 Hz to 50 kHz	
	THD	0.3%	0.35%	Note 1
	IMD	0.12%	0.23%	4:1 mix ratio
	Noise unweighted, (shorted input terminals)	−70 dB	−71 dB	Volume control fully clockwise
Measurements taken at 5 W power output	Input voltage	0.24 V rms	0.24 V rms	Volume control fully clockwise
	Frequency response	±1 dB, 6 Hz to 60 kHz	±1 dB, 7 Hz to 50 kHz	
	THD	0.3%	0.4%	Note 1
	IMD	0.26%	0.44%	4:1 mix ratio
	Noise, unweighted, (shorted input terminals)	−70 dB	−78 dB	Volume control fully clockwise
Measurements taken at 10 W power output	Input voltage	0.35 V rms	0.34 V rms	Volume control fully clockwise
	Frequency response	±1 Hz, 9 Hz to 60 kHz	±1 Hz, 9 Hz to 60 kHz	
	THD	0.6%	0.55%	Note 1
	IMD	0.52%	0.72%	4:1 mix ratio
	Noise, unweighted, (shorted input terminals)	−71 dB	−78 dB	
Measurements taken at 20 W power output	Input voltage	0.46 V rms	0.46 V rms	Volume control fully clockwise
	Frequency response	±1 dB, 12 Hz to 50 kHz	±1 dB, 14 Hz to 50 kHz	
	THD	1.1%	0.9%	Note 1
	IMD	0.69%	1.2%	4:1 mix ratio
	Noise, unweighted, (shorted input terminals)	−75 dB	−80 dB	

(continued)

TABLE 12.11   Measured Performance of the 50 W Stereo Power Amplifier (*Continued*)

Conditions	Parameter	Measured Value, Right Channel	Measured Value, Left Channel	Notes
Measurements taken at 25 W power output	Input voltage	0.52 V rms	0.52 V rms	Volume control fully clockwise
	Frequency response	±1 dB, 14 Hz to 30 kHz	±1 dB, 16 Hz to 50 kHz	
	THD	1.1%	1.2%	Note 1
	IMD	1.1%	1.8%	4:1 mix ratio, note 2
	Noise, unweighted, (shorted input terminals)	−80 dB	−87 dB	

Notes:

1 THD measured at 30 Hz, 100 Hz, 1 kHz, 10 kHz, and 20 kHz unless otherwise noted. The value recorded is the highest reading.

2 The difference in IMD performance between channels is notable. Adjustment of the bias point was optimized for the right channel. The optimal point for the left channel was slightly different. This explains the IMD differences between channels.

and 30 W.[15] Measurements were taken on the right channel amplifier; the left channel performance was essentially identical. The following comments and observations can be made:

- **2 W power output**   1) As with the 12.5 W amplifier, tilt was observed at frequencies of 100 Hz and below. 2) At 1 kHz, response looks quite good. 3) At 10 kHz, a surprising amount of ripple was observed.
- **15 W power output**   1) Low-frequency tilt was observed at 100 Hz, but not unexpected given the measurement at 2 W. 2) The performance at 1 kHz looks quite good—almost picture-perfect. 3) Considerable ripple at 10 kHz was found, but again not unexpected.
- **30 W power output**   1) Low-frequency tilt was observed at 100 Hz, but for an amplifier running at the edge of its capabilities, the performance is not bad. 2) At 1 kHz, with the exception of some minor ringing on the leading edge, the waveform is perfect. 3) The ripple at 10 kHz is not surprising, given the response at lower channels. It is important to note that a fundamental square wave test at 10 kHz will predict amplifier performance with reasonable accuracy from 10 kHz up to 100 kHz or so. No audio source material will approach that limit.

---

[15] Note that this output level is above the rated power output of 25 W. Discussion of the 30 W versus 25 W output capability of this amplifier can be found in Chapter 11. For the square wave measurements, a calculated power output of 30 W was achieved.

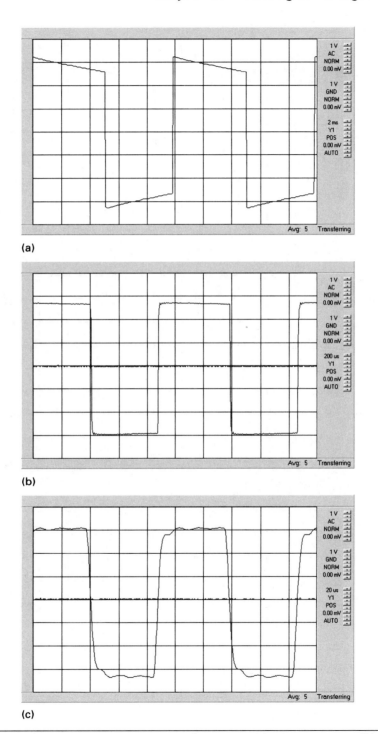

(a)

(b)

(c)

FIGURE 12.15 Square wave response performance: (*a*) 100 Hz at 2 W, (*b*) 1 kHz at 2 W, (*c*) 10 kHz at 2 W (*Continued*)

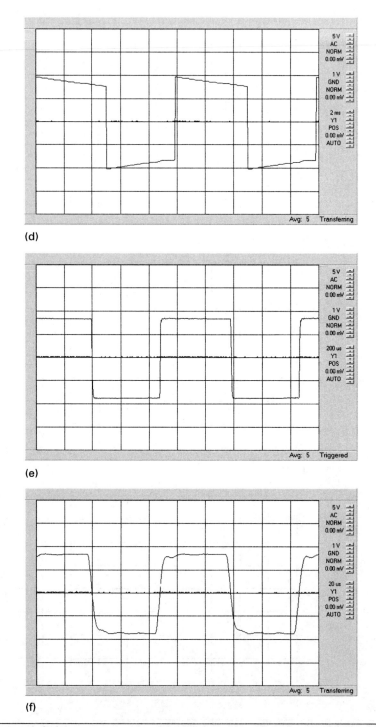

(d)

(e)

(f)

FIGURE 12.15   Square wave response performance: (*d*) 100 Hz at 15 W, (*e*) 1 kHz at 15 W, (*f*) 10 kHz at 15 W (*Continued*)

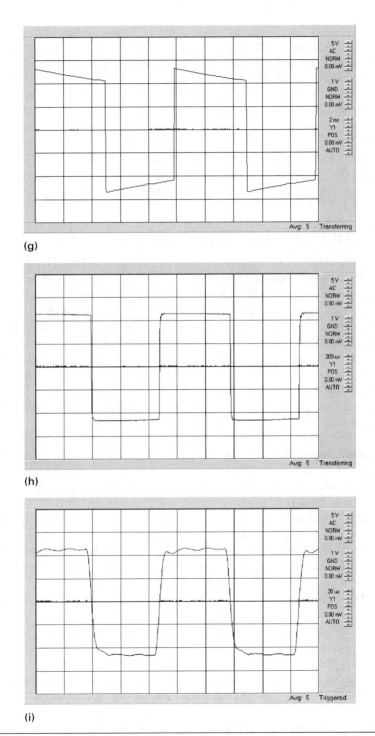

(g)

(h)

(i)

FIGURE 12.15   Square wave response performance: (*g*) 100 Hz at 30 W, (*h*) 1 kHz at 30 W, (*i*) 10 kHz at 30 W

Low-frequency overload is a potential problem in power amplifiers, particularly those with considerable output capability. Overload may be caused by a sudden voltage jump in the input signal, such as a heavy bass passage in program material or connecting (or disconnecting) an input while the power is on. A key consideration in preventing overload problems is to ensure that the amplifier, when overdriven, clips symmetrically. As documented in Chapter 11, this amplifier handles overload quite well. During bench testing, the circuit was routinely driven well into clipping without any observed problems. Lightly loaded and overloaded conditions were also experienced (both intentionally and unintentionally) with no notable issues. Power supply regulation, output transformer design, and careful shaping of the frequency-response curve all contribute to the ability of the amplifier to recover quickly from an overload condition.

About the Specifications

It is important to understand that a cursory review of specifications can lead to an incomplete picture of the characteristics of any audio product. In this book the author has attempted to provide sufficient detail for the reader to judge the overall performance of the circuits presented. Like any product, there are good points and some shortcomings as well. It is expected that minor variations from the stated specifications will be found in any given circuit built by two different people using similar—but not identical—parts. With this preamble, it is useful to judge the performance of a circuit as referenced to some benchmark. Sticking with the Heathkit theme, one of the most respected amplifiers offered by the company was the W-5M amplifier, produced in the mid-1950s. This 25 W mono-block unit still gets good reviews and commands handsome prices on eBay.

Comparing the measurements taken on the 25 W amplifier and the W-5M [1], the following observations can be made:

- The frequency response of the 25 W amplifier compares well with the W-5M across various power levels.
- The harmonic distortion performance of the 25 W amplifier is notably better (lower) than the W-5M at low and high frequencies. Both amplifiers exhibit considerably increased distortion below about 30 Hz. The high-frequency THD performance of the 25 W amplifier is markedly better (lower) than the W-5M at high power levels above about 15 kHz.
- Intermodulation distortion performance of the 25 W amplifier is notably better (lower) than the W-5M at high power levels.
- The W-5M outperforms the 25 W amplifier in hum and noise by more than 10 dB. The stated unweighted noise level of the W-5M relative to 25 W output is –99 dB. Very impressive.
- The square wave characteristics of the 25 W amplifier compare favorably with the W-5M, although only a limited number of tracings are provided by Heathkit, and the power level is not specified (at least not in the 1956 version of the user manual).
- The reserve power of the W-5M is considerably greater than the 25 W amplifier. The stated maximum power output of the W-5M is 47 W, whereas the maximum power output of the 25 W amplifier is about 30 W.

All things considered, the two amplifiers appear to be evenly matched—at least insofar as raw specifications are concerned. The author takes no position on whether one will sound better than the other, as that is a subjective evaluation that each user must make.

Connections

After an amplifier has been completed, it will certainly need to be connected to a pair of speakers. The choices of speaker wire are varied—from common hookup wire ("zip cord") to esoteric (and expensive) cable assemblies. Personal preferences and available cash are the primary determining factors here. The only recommendations the author will make on this subject are as follows:

- Use a twin-lead cable of at least no. 16 gauge.
- Keep the cable length as short as possible to minimize resistive loss and capacitive loading.
- Properly terminate each end of the cable.

An equally wide selection of cables is available commercially to connect source equipment to the preamplifier, and to connect the preamp to the power amp. Keeping with the practice of offering minimum guidelines:

- Use molded cable assemblies. Don't try to make your own unless you really feel the need to; and if so, use high-quality cable and connectors.
- Select cables that have high-quality connectors that fit tightly into the socket. Loose shield connections will lead to hum/noise problems.
- Use cables that are just long enough to do the job. Avoid long cable runs, as this can lead to noise issues.

For audio systems in which the preamplifier is separate from the power amplifier and volume controls are provided on both units, some trial and error may be necessary to find the optimal setting for the power amplifier input in order to minimize noise. Any interconnection of signal lines from one unit to another is best done at the highest possible level, consistent with acceptable distortion performance. The higher the signal levels on the interconnecting cables, the greater the noise immunity. As a first approximation, start by setting the power amplifier input to one-third clockwise rotation and adjust the preamplifier volume control for the desired listening level. Additional adjustments may be needed as more experience is gained with the system.

Although it is probably obvious, it is nonetheless worth stating that all audio equipment should be connected to a source of AC power that is capable of supplying the necessary current for the expected load. Use only grounded outlets. Never defeat the safety ground.

Figure 12.16 shows the stereo preamplifier and 25 W stereo amplifier described in this chapter hooked up and ready to go.

FIGURE 12.16 The finished project

Speakers

The author will provide no guidance on speaker selection, as that falls into another domain entirely. Suffice it to say (please excuse the obvious nature of this advice) that the speakers should be connected to the proper impedance output (4, 8, or 16 Ω), connected with proper polarity observed, and capable of operation at a power level equal to or greater than the rated output of the amplifier.

It is worth noting that trends in speaker usage by consumers have changed over the years, and at the risk of oversimplification, the following general observations are offered for consideration:

- Speaker systems in the 1950s and earlier were a part of the receiver itself, often a large furniture piece.
- The bookshelf-sized speaker was popular in the 1960s and 1970s. Offering good performance and reasonable size, marketing often focused on the construction attributes of the device; for example, a *three-way* system must be better than a *two-way* system, and a *four-way* system was better still. The size of the enclosure varied widely and influenced the performance of the device, since the physics are more easily dealt with when size constraints are lessened.
- During the 1980s, the console stereo had all but disappeared and bookshelf speakers reigned supreme. High-end speaker systems that were furniture pieces in themselves became widely available—and still produce great sound today.
- The 1990s marked a turning point for speakers, where the general consensus seemed to be moving toward a belief that bigger was not necessarily better. Speaker systems became smaller with the separation of the bass transducer from the left and right channel "speakers." This was an ingenious solution to the

very real problem of 5.1-channel surround sound. Most family rooms would be challenged to place (for optimum listening) five separate bookshelf speakers of moderate size. As smaller became better, some consumers (I know of several) wanted the speakers to disappear completely—but not the sound. This trend led to recessed and otherwise hidden speakers, often with some sonic compromise.

Where are things today? The small multichannel system is still very popular, and the size benefit of this approach is even more important as content producers begin rolling out 7.1-channel surround sound offerings. Interestingly, however, it seems that the bookshelf speaker is making a bit of a comeback, at least for audio listening. The big, specialized, high-end speaker systems are still being purchased and enjoyed by those who have the means to afford them, and the space to showcase them.

Duty Cycle

While it is possible to leave a vacuum tube audio amplifier on for days at a time, it is usually not recommended. Because tubes have a finite life, audio gear should be powered on only when it is being used. The possibility of component failure is another consideration.

It was common practice decades ago to switch things off when they were not in use. That practice largely faded from the scene, making a return only recently because of energy-conservation concerns.

A vacuum tube audio amplifier is different from a solid-state amplifier (the author would argue it is not just different, but better). A return to "appointment listening" would be a welcomed change.

Build or Rebuild?

The W-5M performance comparison discussion brings up another important point, and a suitable one to close on—namely, build from new or rebuild? A number of excellent classic amplifiers are available on eBay and from other sources at reasonable (and sometimes unreasonable) prices. It is certainly possible to purchase an old, perhaps inoperative, unit and return it to mint condition. This is a viable option for many people. The benefits include: 1) the amplifier is a proven design, 2) there is usually no chassis work (other than cleaning/painting) required, and 3) the process is simpler and usually faster. The cost of the rebuild approach is difficult to assess, since the prices for vintage classic products vary widely.

Building an amplifier new from scratch has its benefits as well, including: 1) all-new construction means greater long-term reliability (or at least it should), 2) the unit can be designed and personalized as desired, 3) it is truly in "mint condition" since it is brand-new, and 4) there is considerable satisfaction in building something up from a pile of parts.

So, the short answer to the question of which way to go—build or rebuild—is that it depends on what you want. Either way, the fundamental principles of vacuum tube operation and good construction practices apply.

In the realm of vacuum tube audio there is no shortage of opinions on what constitutes a good amplifier. Some of the discussion is quite intricate and detailed. For the hobbyist, however, the only things that really matter are whether the builder is satisfied with the final outcome, whether the project was an enjoyable one, and whether it came in somewhere close to budget.

For those looking for the ultimate audio system, the options are numerous and easily found. Just add money. For the rest of us, building a good audio system is less demanding, and hopefully more fun.

Chapter 13

Final Thoughts

In the world of technical standards development, a profession that I know well (my day job), the project chapters of this book could be described as a *travelogue*. A travelogue, you see, reads something like the following: "We did this, and then this, and then this, and finally wound up here." For a technical standards document, it can be a distraction to describe the steps to the final decision—it is better to just state the final technical solution.

Having said all that, in preparing this book it felt right to include details of each step in the process, which can be summarized as follows:

1. Select an interesting and promising circuit from a classic source. *Check. Done.*
2. Attempt to find the needed parts. *Check.* Mostly successful, with some compromises along the way.
3. Build the circuit and measure the performance. *Check.* The results were mostly as expected, but there were a few surprises along the way.
4. Try some modifications to the basic circuit that seem promising or that solve a problem experienced with the initial build. *Check.* In some cases, considerable improvements were identified and implemented; for example, the screen tap option described in Chapters 10 and 11.
5. Arrive at a final circuit design and final parts list. *Check.* Developing the parts lists the second time is far easier than the first.
6. Develop a PWB layout for the final design. *Check.* In this case, the second time was a charm. The first PWB versions worked but were not perfect.[1]
7. Put all of the elements together into a finished product that works well and looks good. *Check.*
8. Put on Sinatra and enjoy the great-sounding audio. *Check* and *check.*

I think that the travelogue approach works here. I hope you agree. As a practical matter, no design is perfect, and if left to the designer, a product might never be finished because additional improvements could likely be found. The intent of the

---

[1] In design work, Rev. 2 is always better than Rev. 1.

travelogue approach is to demonstrate a series of proven steps that engineers—and hobbyists—have used for a very long time. Ideas for circuit improvements can come from careful study, suggestions from others working on similar projects, or just a hunch. Some improvements yield exceptional results; others not so much.

In the preface, I mentioned the Heathkit experience that many folks of a certain age[2] enjoyed years ago. I dare say that most kit builders remember more about building the kits than about using them. Building something up from a pile of parts is a wonderful experience, particularly when it is something you can put in your living room and enjoy—like audio gear.

I see this book as a starting point. It is my hope that another book will follow, but that's a whole different discussion involving various nice folks who crunch numbers and come up with promising ideas. Regardless, I intend to keep going, if only with a web presence.

Future projects queued up with parts standing by include more amplifiers, receivers, and maybe even some audio test equipment. As with this book, these projects will begin with a classic design and go from there, providing travelogue commentary along the way.

So, no shortage of projects—limited only by available time and money.

More to come...

—*Jerry C. Whitaker*

---

[2] I really don't like that expression, since now I am now "of a certain age."

VacuumTubeAudio.info

The author has established a website to support this book and other planned projects that build on it. The following information can be found at www.VacuumTubeAudio.info:

- Parts lists for the projects described in this book as downloadable Microsoft Word or Excel files
- PWB layout resources
- Additional supporting information and commentary on the projects contained in this book
- Additional audio projects using vacuum tubes, including receivers and test equipment
- Kits and components available for purchase
- Vacuum tube audio gear available for purchase
- Other interesting stuff relating to vacuum tube audio applications

Please visit www.vacuumtubeaudio.info and see what's new.

Appendix

Notes and References

Chapter 1

[1] *RCA Receiving Tube Manual*, RC-30, Radio Corporation of America, Camden, NJ, 1975.

[2] Benson, K. Blair: "Electronic Engineering Fundamentals," in *Television and Audio Handbook for Technicians and Engineers*, K. Blair Benson and Jerry C. Whitaker (eds.), McGraw-Hill, NY, 1990.

[3] Fink, Donald G., and Don Christiansen (eds.): *Electronic Engineers' Handbook*, McGraw-Hill, New York, NY, 1982.

[4] Cabot, Richard C.: "Audio Test and Measurements," in *Audio Engineering Handbook*, K. Blair Benson (ed.), McGraw-Hill, New York, NY, 1988.

[5] Toole, Floyd E., E. A. G. Shaw, G. A. Daigle, and M. R. Stinson: "The Physical Nature of Hearing," in *Standard Handbook of Audio and Radio Engineering*, 2nd edition, Jerry C. Whitaker and K. Blair Benson (eds.), McGraw-Hill, New York, NY, 2002.

Bibliography

Benson, K. Blair, and Jerry C. Whitaker: *Television and Audio Handbook for Technicians and Engineers*, McGraw-Hill, New York, NY, 1990.

Benson, K. Blair: *Audio Engineering Handbook*, McGraw-Hill, New York, NY, 1988.

Rhode, U., J. Whitaker, and T. Bucher: *Communications Receivers*, 2nd ed., McGraw-Hill, New York, NY, 1996.

Whitaker, Jerry C., and K. Blair Benson (eds): *Standard Handbook of Video and Television Engineering*, McGraw-Hill, New York, NY, 2000.

Whitaker, Jerry C.: *Electronic Systems Maintenance Handbook*, 2nd Ed., CRC Press, Boca Raton, FL, 2002.

Whitaker, Jerry C.: *Video and Television Engineer's Field Manual*, McGraw-Hill, New York, NY, 2000.

Chapter 2

[1] Benson, K. Blair: "Components and Typical Circuits," in *Television and Audio Handbook for Technicians and Engineers*, K. Blair Benson and Jerry C. Whitaker (eds.), McGraw-Hill, NY, 1990.

[2] Whitaker, Jerry C.: *The Electronics Handbook*, CRC Press, Boca Raton, FL, 1996.

[3] Benson, K. Blair, and Jerry C. Whitaker: *Television and Audio Handbook for Technicians and Engineers*, McGraw-Hill, New York, NY, 1990.

[4] Whitaker, Jerry C.: *AC Power Systems Handbook*, 3rd. ed., CRC Press, Boca Raton, FL, 2007.

[5] Heathkit: "FET Multimeter IM-5225 Manual," Heath Company, Benton Harbor, MI, 1977.

[6] Parker, Martin R., and William E. Webb: "Magnetic Materials for Inductive Processes," in *The Electronics Handbook*, 2nd ed., Jerry C. Whitaker (ed.), CRC Press, Boca Raton, FL, 2005.

[7] Whitaker, Jerry C., Eugene DeSantis, C. Robert Paulson: *Interconnecting Electronic Systems*, CRC Press, Boca Raton, FL, 1992.

[8] Whitaker, Jerry C.: *Power Vacuum Tubes Handbook*, 3rd ed., CRC Press, Boca Raton, FL, 2011.

Bibliography

Benson, K. Blair: *Audio Engineering Handbook*, McGraw-Hill, New York, NY, 1988.

Fink, Donald G., and Don Christiansen (eds.): *Electronic Engineers' Handbook*, McGraw-Hill, New York, NY, 1982.

Rhode, U., J. Whitaker, and T. Bucher: *Communications Receivers*, 2nd ed., McGraw-Hill, New York, NY, 1996.

Whitaker, Jerry C.: *Electronic Systems Maintenance Handbook*, 2nd ed., CRC Press, Boca Raton, FL, 2001.

Whitaker, Jerry C.: *Video and Television Engineer's Field Manual*, McGraw-Hill, New York, NY, 2000.

Whitaker, Jerry C., and K. Blair Benson (eds): *Standard Handbook of Video and Television Engineering*, McGraw-Hill, New York, NY, 2000.

Chapter 3

[1] *RCA Receiving Tube Manual*, RC-30, Radio Corporation of America, Camden, NJ, 1975.

[2] Whitaker, Jerry C.: *Power Vacuum Tubes Handbook*, 3rd ed., CRC Press, Boca Raton, FL, 2011.

[3] Terman, F. E.: *Radio Engineering*, 3rd ed., McGraw-Hill, New York, NY, 1947.

[4] Ferris, Clifford D.: "Electron Tube Fundamentals," in *The Electronics Handbook*, Jerry C. Whitaker (ed.), CRC Press, Boca Raton, FL, pp. 295–305, 1996.

[5] Laboratory Staff: *The Care and Feeding of Power Grid Tubes*, Varian Eimac, San Carlos, CA, 1984.

[6] Spangenberg, Karl R.: *Vacuum Tubes*, McGraw-Hill, New York, NY, 1948.

Bibliography

Chaffee, E. L.: *Theory of Thermonic Vacuum Tubes*, McGraw-Hill, New York, NY, 1939.

Eastman, Austin V.: *Fundamentals of Vacuum Tubes*, McGraw-Hill, New York, NY, 1941.

Fink, D., and D. Christiansen (eds.): *Electronics Engineers' Handbook*, 3rd ed., McGraw-Hill, NY, 1989.

Harper, C. A.: *Electronic Packaging and Interconnection Handbook*, McGraw-Hill, NY, 1991.

Jordan, Edward C., (ed.): *Reference Data for Engineers: Radio, Electronics, Computer and Communications*, 7th ed., Howard W. Sams, Indianapolis, IN, 1985.

Kohl, Walter: *Materials Technology for Electron Tubes*, Reinhold, NY.

Reich, Herbert J.: *Theory and Application of Electronic Tubes*, McGraw-Hill, New York, NY, 1939.

Whitaker, J. C.: *Radio Frequency Transmission Systems: Design and Operation*, McGraw-Hill, New York, NY, 1991.

Chapter 4

[1] Whitaker, Jerry C.: *Power Vacuum Tubes Handbook*, 3rd ed., CRC Press, Boca Raton, FL, 2011.

[2] *RCA Receiving Tube Manual*, RC-30, Radio Corporation of America, Camden, NJ, 1975.

[3] Laboratory Staff: *The Care and Feeding of Power Grid Tubes*, Varian Eimac, San Carlos, CA, 1984.

[4] Terman, F. E.: *Radio Engineering*, 3rd. ed., McGraw-Hill, New York, NY, pg. 560, 1947.

[5] Douglass, Barry G.: "Thermal Noise and Other Circuit Noise," in *The Electronics Handbook*, Jerry C. Whitaker (ed.), CRC Press, Boca Raton, FL, pp. 30–36, 1996.

Bibliography

Crutchfield, E. B. (ed.): *NAB Engineering Handbook*, 8th ed., National Association of Broadcasters, Washington, DC, 1992.

Jordan, Edward C. (ed.): *Reference Data for Engineers: Radio, Electronics, Computers, and Communications*, 7th ed., Howard W. Sams, Indianapolis, IN, 1985.

Whitaker, Jerry C.: *AC Power Systems*, 2nd ed., CRC Press, Boca Raton, FL, 1998.

Whitaker, Jerry C.: *Radio Frequency Transmission Systems: Design and Operation*, McGraw-Hill, New York, NY, 1991.

Chapter 5

[1]   Cabot, Richard C.: "Audio Test and Measurements," in *Audio Engineering Handbook*, K. Blair Benson (ed.), McGraw-Hill, New York, NY, 1988.

[2]   Laboratory Staff: *The Care and Feeding of Power Grid Tubes*, Varian Associates, San Carlos, CA, 1984.

[3]   Besch, David F.: "Thermal Properties," in *The Electronics Handbook*, Jerry C. Whitaker (ed.), CRC Press, Boca Raton, FL, pp. 127–134, 1996.

[4]   Whitaker, Jerry C.: *Power Vacuum Tubes Handbook*, 3rd ed., CRC Press, Boca Raton, FL, 2011.

[5]   *RCA Receiving Tube Manual*, RC-30, Radio Corporation of America, Camden, NJ, 1975.

Bibliography

Benson, K. B., and J. Whitaker: *Television and Audio Handbook for Engineers and Technicians*, McGraw-Hill, New York, NY, 1989.

Lanphere, John: "Establishing a Clean Ground," *Sound & Video Contractor*, Intertec Publishing, Overland Park, KS, August 1987.

Mullinack, Howard G.: "Grounding for Safety and Performance," *Broadcast Engineering*, Intertec Publishing, Overland Park, KS, October 1986.

Rising, Roy: "Audio Interconnection," in *The Electronics Handbook*, Jerry C. Whitaker (ed.), CRC Press, Boca Raton, FL, 1996.

Tremaine, H. M.: "Installation Techniques," in *Audio Cyclopedia*, 2nd ed., Howard W. Sams & Co., Indianapolis, IN, 1969.

Whitaker, Jerry C., Eugene DeSantis, C. Robert Paulson: *Interconnecting Electronic Systems*, CRC Press, Boca Raton, FL, 1992.

Whitaker, Jerry: *Maintaining Electronic Systems*, CRC Press, Boca Raton, FL, 1991.

Chapter 6

[1]   "Tips for Designing PCBs," ExpressPCB, www.expresspcb.com/ExpressPCBHtm/Tips.htm

[2]   Whitaker, Jerry C.: *Maintaining Electronics Systems*, CRC Press, Boca Raton, FL, 1991.

[3]   Heathkit: "Harmonic Distortion Analyzer," Heath Company, Benton Harbor, MI, 1976.

[4]   Heathkit: "FET Multimeter IM-5225 Manual," Heath Company, Benton Harbor, MI, 1977.

[5]   Whitaker, Jerry C.: *Power Vacuum Tubes Handbook*, 2nd ed., CRC Press, Boca Raton, FL, 2003.

[6]   Cabot, Richard C., and Robert Metzler: "Nonlinear Audio Distortion," in *Standard Handbook of Broadcast Engineering*, Jerry C. Whitaker (ed.), McGraw-Hill, New York, NY, 2005.

[7] Hoyer, Mike: "Bandwidth and Rise Time: Two Keys to Selecting the Right Oscilloscope," *Electronic Servicing and Technology*, Intertec Publishing, Overland Park, KS, April 1990.

[8] Heathkit: "IT-17 Tube Checker," Heath Company, Benton Harbor, MI.

Bibliography

Bausel, James: "Focus on Soldering and Desoldering," *Electronic Servicing and Technology*, Intertec Publishing, Overland Park, KS, November 1989.

Fenton, Christopher: "Choosing a Soldering Iron," *Electronic Servicing and Technology*, Intertec Publishing, Overland Park, KS, May 1988.

Graham, Edward S.: "Designing a Working Service Kit," *Microservice Management*, Intertec Publishing, Overland Park, KS, May 1989.

O'Brien, Gil: "Cleaning Supplies for Computers," *Microservice Management*, Intertec Publishing, Overland Park, KS, August 1988.

OSHA: "Electrical Hazard Fact Sheets," U.S. Department of Labor, Washington, DC, January 1987.

Persson, Conrad: "I Only Have Two Hands," *Electronic Servicing and Technology*, Intertec Publishing, Overland Park, KS, July 1986.

Persson, Conrad: "Locating Replacement Parts", *Electronic Servicing and Technology*, Intertec Publishing, Overland Park, KS, December 1987.

Persson, Conrad: "Setting Up a Test Bench," *Electronic Servicing and Technology*, Intertec Publishing, Overland Park, KS, March 1987.

Persson, Conrad: "Solder: The Tin That Binds," *Electronic Servicing and Technology*, Intertec Publishing, Overland Park, KS, February 1986.

Richardson, Jeff: "The Benefits of a Tool Kit Program," *Microservice Management*, Intertec Publishing, Overland Park, KS, May 1989.

Smeltzer, Dennis: "Packing and Shipping Equipment Properly," *Microservice Management Magazine*, Intertec Publishing, Overland Park, KS, April 1989.

Whitaker, Jerry C., G. DeSantis, and C. Paulson: *Interconnecting Electronic Systems*, CRC Press, Boca Raton, FL, 1993.

Whitaker, Jerry C.: *AC Power Systems*, 3rd ed., CRC Press, Boca Raton, FL, 2004.

Whitaker, Jerry C.: *Maintaining Electronic Systems*, CRC Press, Boca Raton, FL, 1991.

Whitaker, Jerry C.: *Radio Frequency Transmission Systems: Design and Operation*, McGraw-Hill, New York, NY, 1990.

Chapter 7

[1] *RCA Receiving Tube Manual*, RC-30, Radio Corporation of America, Camden, NJ, 1975.

Chapter 8

[1] *RCA Receiving Tube Manual*, RC-30, Radio Corporation of America, Camden, NJ, 1975.

Chapter 9

[1]   *RCA Receiving Tube Manual*, RC-30, Radio Corporation of America, Camden, NJ, 1975.

[2]   Benson, K. Blair, *Audio Engineering Handbook*, McGraw-Hill, New York, NY, 1988.

Chapter 10

[1]   *RCA Receiving Tube Manual*, RC-30, Radio Corporation of America, Camden, NJ, 1975.

Chapter 11

[1]   *RCA Receiving Tube Manual*, RC-30, Radio Corporation of America, Camden, NJ, 1975.

Chapter 12

[1]   *Heathkit High-Fidelity Amplifier Model W-5M*, Heath Company, Benton Harbor, MI, 1956.

Index

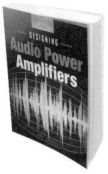